Money-Making 900 Numbers

How Entrepreneurs Use the Telephone to Sell Information

Written by:

Carol Morse Ginsburg

and

Robert Mastin

Aegis Publishing Group, Ltd.
796 Aquidneck Avenue
Newport, Rhode Island 02842
401-849-4200

Library of Congress Catalog Card Number: 94-74497

International Standard Book Number: 0-9632790-1-7

Printed in the United States of America.
First Edition, first printing, 1995

10 9 8 7 6 5 4 3 2 1

This publication is designed to provide accurate and authoritative information in regard to the subject matter covered. It is sold with the understanding that neither the author nor the publisher are engaged in rendering legal, accounting or other professional service. If legal advice or other expert assistance is required, the services of a competent professional person should be sought.

Publisher's Cataloging In Publication Data
Ginsburg, Carol Morse and Mastin, Robert L.
Money-Making 900 Numbers: How Entrepreneurs Use the Telephone to Sell Information / by Carol Morse Ginsburg and Robert Mastin.
Includes Index
1. Audiotex services industry.
HE8817.M3 1994   384.64   94-74497
ISBN 0-9632790-1-7

Acknowledgements

Carol Morse Ginsburg:

This book could never have been written without the assistance of Debra A. Velsmid, the assistant editor at *Audiotex News*, and Kimberly Tennant, production assistant, and Carol Weiss, our copy editor, who calmly and faithfully with nary a complaint edited the original material, and of course Mrs. Haggerty, who managed the office so well there was the time and space to write.

To Sherwood, my incomparable husband, who nagged and cajoled me to finish and never gave up on me or the project. To my dear children, Alison, Mary, Eileen Doug and Hal, who always understood, excused, believed and supported me.

And finally to my telephone friends -- too numerous to mention -- who have spent countless hours over the last five years talking, counselling, speculating, explaining and providing countless insights about the pay-per-call industry.

And of course, my co-author Robert Mastin, who was truly a delight to work with!

Robert Mastin:

This book was a collaborative effort with the help of many people. Although it would be impossible to mention everyone who contributed materials, advice or assistance to this work, I would like to express my appreciation to all the participants in the in-depth profiles for their time and participation, which was always cheerfully given despite some very busy schedules.

Thanks to Kimlee Revock and Elaine Anderson for helping maintain some semblance of sanity around the office. A special thanks to John Robinson at ImageTek for an outstanding cover design. And my lovely wife, Liz, deserves a medal for not only putting up with me, but for always giving encouragement when I need it the most.

And finally, my co-author Carol, who was willing to share credit for work that is predominantly hers, as the basis for this book is from five years of her work putting out *Audiotex News* every month without interruption.

Contents

Resource Guide:

Foreword

The authors of this book, Ginsburg and Mastin, feel passionately about this industry -- its vast potential and its infinite possibilities. Their book may not be all there is to know (for that's impossible to compile), but it is what you *need* to know -- all the critical lessons are here.

In on this modern-age business from the very beginning, these authors embody that rare combination -- entrepreneurial excitement tempered by pragmatism. Their feet stand firmly on the ground -- they witnessed the trials and failures as well as the successes -- while they commit themselves to helping the good new ideas succeed. And they're the proven experts. Carol Morse Ginsburg's monthly newsletter, *Audiotex News*, goes well beyond mere description of new applications; she presents up-to-the-minute news about industry development and personalities, plus insightful investigative reporting about the tangled legal and regulatory environment. Robert Mastin's *900 Know-How* enjoys a well-deserved reputation as the bible for new audiotex entrepreneurs.

The development of the pay-per-call industry in the late 1980's can be seen as a largely symbolic event -- one that typified that decade in which the world's largest civilian employer, AT&T, was forced by the courts to shatter into literally hundreds of pieces. It was the decade in which a place was reserved at the table of American business for the entrepreneur.

By breaking up AT&T, the courts sought to foster a new era of competition in the information and telecommunications marketplace. This opened the door for entrepreneurs to create a variety of new voice information services, and because the telephone companies did not want to be left out of this new opportunity, they began to offer billing and collection services to support burgeoning new information products.

Although pay-per-call now exists all over the world, it is largely a product of the uniquely American entrepreneurial experience -- an experience parallel to our history and folklore. For better and sometimes for worse, we have our "homesteaders," who have staked-out a business for the long haul. We have the "gunslingers," who "shot the place up" for a quick buck early on and by now are chasing the next exploitation opportunity. We have our "adventurers," who constantly push the envelopes of new technologies and marketplace needs, and we have our "deputies," with and without official status, who seek to keep the peace in our pioneer community. To say that we attract diversity would be an understatement -- we have the most "buttoned-down" corporate characters, and some you might think Damon Runyon dreamed up, and everything between. Significantly, a large number of women and other business-world minorities hold positions of leadership.

We form a bridge between more traditional manufacturing-oriented industries and the new information/service-oriented ones. Because the telephone is almost everywhere and people feel comfortable using it, audiotex services have met with ready consumer acceptance. But make no mistake, ours is an industry which is setting the precedents that many other sectors of the information industry will have to live by. Optical and on-line services, for example, are only now beginning to address some of the business and social issues created by the information revolution. These technologies will vastly improve the way we live, the way we educate, edify and entertain ourselves. But they will also force us to address the critical issues of our times: access -- because timely information is bound to become more important; privacy -- because information cannot be repossessed; and intellectual property rights -- because the media will be democratized as everyone participates in creating and manipulating content.

If this sound lofty, it's not. Like any business the purpose is to make money. But unlike many businesses, the rules of engagement are still being written. Models that effectively answer such questions as what is the role of advertising?; what is the relative value of raw material content in this case, and the finished product?; and what is a fair cost for distribution?; have yet to be fully developed. That is why this book is so important, particularly to the pay-per-call entrepreneur.

The many creative applications described herein will intrigue and inspire you. Be sure you read between the lines (and note the commentaries and the asides) for an exclusive inside look at the personalities and the relationships among the carriers, service bureaus and information providers, and how they interact. See first hand how new business models are being developed for an entirely new business.

Prospective service providers will find the comprehensive back-of-book resources indispensable. You'll find out how public policy relates to business on the information frontier, the importance of trade shows and industry groups in providing networking opportunities, and all the resources you'll need for success now and into the future.

If you decide that this is the business for you, we welcome you! Ginsburg and Mastin are sterling examples of what a collegiate industry this can be. Competitors become friends, working together not only to build a profitable industry (a rising tide lifts all boats) but to demonstrate the importance of value to our consumers. Voice technology is enhancing the quality of life by improving the way we interact with each other and with traditional media. Into the 21st century, our civilization will progress along this path, providing golden opportunities for the bold and enlightened entrepreneur to achieve financial success.

Peter J. Brennan
Boston, Massachusetts

Peter J. Brennan is Director of Development for Tele-Publishing, Inc. of Boston, a leading service bureau which counts among its clients more than three hundred and fifty daily, weekly and monthly newspapers and magazines in the United States and Canada, as well as a variety of national and local media groups. Brennan was instrumental in gorging a variety of audiotext media partnerships which have become an industry mainstay.

Mr. Brennan is a frequent public speaker and published expert in the field of voice services, technology, legislation and applications, and has offered testimony before both houses of the US Congress, various Attorneys General and Public Utility

Commissions. In 1992 he received the Information Industry Association Distinguished Service Award for leadership as Chair of IIA's Voice Information Services Division. He currently Chairs the Interactive Telephone Special Interest Group of the Interactive Services Association.

Chapter 1
Introduction (Preamble)

Most people think of the information superhighway as some distant vision of the future, with sophisticated high-tech computer networks which are way beyond the comprehension of ordinary people. Not so. The information superhighway is here today in the form of telephone networks, cable TV systems, satellite links, cellular networks and fiber optic cable transmission systems.

We have all been travelling this highway whenever we pick up the telephone to get whatever information we need. Telephone-based information services are but one lane on the information superhighway, though admittedly one of the middle lanes. The fast lane is currently claimed by on-line computer services such as CompuServe and Prodigy, accessible only to those with fast cars and the ability to drive them.

According to Link Resources, a market research firm based in New York, only 30 percent of the 96.3 million U.S. households have a PC. And most of these are used for games and entertainment, not for sophisticated communications and data retrieval using the fast lane of the information superhighway. That leaves 67 million households without PCs of any kind, and no access to the fast lane. People with PCs and the skills to put them to good use aren't the only ones who need specialized information. We can all benefit from access to information that helps make our lives more convenient, productive and enjoyable: stock and commodity prices, airline flight information, technical or professional assistance, movie reviews, sports scores, ski conditions, car pricing information and so on.

One reason for the rapid proliferation of telephone-based information services was the introduction of 900 number pay-per-call services in 1987. A 900 number allows the information provider (IP) to charge a fee in excess of the basic telephone transport charge in order to cover the value of the information content of the program. The telephone company then bills the caller on behalf of the IP and sends the IP's share of the call revenue after holding back a percentage for its services and for future chargebacks. This is obviously a very convenient arrangement for all parties to the transaction. The caller doesn't have to bother writing a separate check and needn't have a credit card handy. The information provider doesn't have to prepare separate bills for potentially thousands of customers for a fairly small sum of money.

In its simplest form 900 pay-per-call is a billing mechanism. It was the entrepreneur's ability to make a profit by offering information services over the telephone that resulted in a virtual explosion of information offerings. The entrepreneur would never have been able to earn a reasonable profit if he had to perform billing and collections for many small transactions. The premium billing services provided by the telephone companies made it possible for entrepreneurs of any size to sell information cost-effectively over the telephone.

Interactive voice processing has merged with 900 number information services to open up a new lane on the information superhighway for people who do not have access to PCs. The telephone keypad becomes the equivalent of the computer keyboard, and any telephone owner has potential access to the same wide array of database information available to on-line computer service subscribers. A lot more people have telephones than have computers, and virtually all of them know how to operate their telephones with relative competence. Imaginative entrepreneurs are responding to the challenge by providing an ever expanding array of information services, accessible to anyone with a telephone.

Will computers eventually replace the basic telephone? After all, many people are hanging their hats on computer-telephone integration (CTI) and computer telephony, new buzz terms you'll hear more and more about. Well, nobody really knows exactly where we're headed, and trying to predict the future is somewhat risky given the rapid pace of technological advancement in computers and telecommunications. Whatever happens, it won't happen overnight. A reasonably powerful computer, with the necessary peripherals and software, still costs around $2,000, while the lowly telephone can be purchased for $20 or less. It is probably a pretty safe bet that the basic telephone will be around for many more years to come.

This book is intended for individuals or businesses who wish to get started in the business of selling information by telephone. In essence, this is an idea book. You will read about all kinds of information services being sold over 900 numbers. Many of these programs were short-lived and are no longer functional. Others have been strong, long-term performers that have been in operation for years.

The Profiles

Nearly all the programs profiled here come from the pages of past issues of *Audiotex News*, the preeminent trade newsletter for the 900-number pay-per-call industry since 1988. Both functional as well as non-functional programs are included here.

Wherever possible, an attempt was made before going to press to establish the current status of each program, whether or not it was still operational. Operational programs that could still be reached via the published 900 number are denoted with "*(active)*." The absence of any such designation means that we could not establish that the program was still operational, or that the current operational status is irrelevant (one-time event programs).

For the most part, the status of each program was verified by simply calling the 900 number that is listed. As we will explain later, the absence of an "active" designation does not necessarily mean that the program itself is no longer operational. It simply means that the program is no longer accessible on that particular 900 number. Nevertheless, most of such programs are indeed defunct, but we have no way of conclusively ascertaining which ones moved to another number and which ones have ceased to exist altogether.

Why include non-operational programs? Because learning about the failures can be more enlightening than hearing about nothing but success stories. You cannot be exposed to too much information when evaluating the feasibility of launching a particular 900-number service. You should know about the programs that didn't make it along with those that did. Only then will you be able to make an informed decision regarding the prospects for your program application.

There are many inactive programs listed here. Don't despair if you find some in the subject area that interests you, because the reason for the program no longer being active isn't necessarily because it was a failure or a bad idea. It could have been an excellent program, with a well-defined market, but perhaps the IP simply didn't have sufficient capital to hang in long enough to see it succeed. This is one of the most common reasons for the failure of many businesses.

The program may have been on one of Telesphere's lines, a long distance 900 number carrier that abruptly went out of business back in September 1991, putting quite a few unsuspecting IPs out of business, through no fault of their own. Many of these IPs had a lot of money tied up in advertising their 900 numbers, and with no warning the plug was pulled and all calls to Telesphere 900 numbers went unanswered. It was very difficult for many of these small entrepreneurs to recover from such a serious financial setback.

Some programs are purposely designed for a one-shot event or happening, with no intention of ever being an ongoing program. Programs relating to the Olympics, Word Cup Soccer, or music concert tours are examples of this type application.

A program could be terminated because the IP decided to move on into other areas despite the fact that it was profitable. Or, it may have been a great idea but it just wasn't ever executed very well. Maybe the IP decided to retire, or perhaps his career has been interrupted by an indeterminate stay at the local correctional facility. Or, the program was never marketed properly to the right target audience.

The information provider may have changed service bureaus or long distance carriers, and is now using a different 900 number. As of this writing, 900 numbers are not transportable, so changing carriers means changing numbers.

The 900 number may be only regional, or the IP may have blocked access from the east coast, where we were calling from to check on operational status. IPs may have many reasons for blocking access from a given geographic area, such as an overabundance of deadbeats or market research that demonstrates that the area is a poor market for the particular service.

The line could be only seasonal in nature, such as lines dealing with ski conditions or NFL scores.

The point is, there could be a multitude of reasons for a program being defunct, so we shouldn't be too quick to judge all non-active programs as failures or as bad ideas. The bottom line is, we thought it would be helpful to present both the non-operational along with the operational, and let you be the judge of whether or not the program was a good idea to begin with.

Another piece of useful information included with each profile is the issue date the write-up first appeared in *Audiotex News*. This is particularly helpful for the operational programs, giving you an indication of the longevity of the program. Obviously, long-lived still-operational programs are quite successful by any standard.

In addition to the numerous brief profiles drawn from past issues of *Audiotex News*, several in-depth profiles are presented for some of the more interesting applications. Here we tried to get into the reasons behind initiating the program, how it is marketed, to whom, its level of success, and so on. Trying to get concrete financial data from IPs, however, is kind of like trying to get atomic secrets from the Pentagon. Successful businesses in particular are understandably hesitant about divulging the secrets to their success, lest their competitors are given the business intelligence necessary to catch up. Nonetheless, we dug as deep as we could, knowing that you would be most interested in concrete details about how much money can be made with a given program. If this information is missing it isn't for lack of trying to get it.

Don't let the number of profiles in any given category lead you to any conclusions about the viability or profitability of that particular type of program. This is nothing more than a broad cross-section of information and entertainment programs.

In fact, the most prevalent -- and the most profitable -- 900 number programs have historically been the adult lines and the dating lines. Well, there is only so much that can be said about either application because they're all pretty much the same.

The purpose here is not to be a complete listing of all 900 programs that ever existed. There are too many -- around 10 thousand is the conventional wisdom -- and the book would get very repetitive and quite boring. This is a representative cross-section of what's out there, past and present. We apologize if we missed a particularly interesting application. If you know of one that belongs in a future edition of this book please send complete details to Aegis Publishing Group or to *Audiotex News*.

Because these *Audiotex News* profiles start shortly after the birth of this industry, beginning in 1990, this book also serves as a historical record of this business from its infancy to its current mature status. You will get a real appreciation for how the business has progressed from the entrepreneurial wild west days to a more bottom-line oriented established industry dominated by savvy marketers and large companies.

Should this book motivate you to jump into this exciting business, check out the appendices at the end of the book. You will find comprehensive resources and useful references for further research into this business.

What is 900?

Because most telephone-based pay-per-call programs are offered over 900 numbers, it is necessary at this point to discuss what this industry is all about. First of all, calling it an industry is a bit of a stretch, but we'll keep doing it anyway in the interest of brevity, instead of calling it "The 900 pay-per-call information delivery service" or by some other equally unwieldy title.

A 900 number is simply an alternative method of paying for information over the telephone. It is nothing more or less than a convenient information delivery medium, with a very efficient way of exchanging payment for information delivered. Whether the charge is by the minute or a flat rate for the call, the caller is charged for the telephone call on his or her monthly phone bill. Basically, the reverse of an 800 number. The person offering the information or entertainment, the information provider (IP), has the latitude of charging whatever the market will bear, within some fairly generous limits imposed by the long distance telephone carrier. You will also see the terms "pay-per-call," "976," and "caller-paid" used interchangeably along with simply, "900." They all mean essentially the same thing: the caller is charged for the telephone call, at a rate in excess of normal toll charges, in exchange for the information or entertainment services provided.

The 900 industry was launched in 1980 by AT&T, then calling it "DIAL-IT 900 Service," with the premier of DIAL-IT National Sports (see chapter 7 in-depth profile of Phone Programs, Inc.) in September, followed shortly thereafter in October by ABC-TV's use of the service during the Reagan-Carter presidential debates to poll viewer opinion. On the final night of the debates, 500,000 people paid 50 cents each to register their opinions on who won the debate.

Maybe you remember the *Saturday Night Live* episode in 1982 when viewers were asked to call a 900 number to vote whether or not Eddie Murphy should boil Larry the Lobster. A whopping 500,000 callers participated, voting to save poor Larry, but Eddie boiled him anyway.

The industry really didn't take off, however, until 1987 when AT&T began offering premium billing services. This allowed IPs to generate profits from calls instead of simply covering the program costs. During the same year, Telesphere initiated the first interactive 900 service. By 1989, the three major long distance carriers (AT&T, MCI and Sprint) were offering interactive 900 services, and industry revenues were growing very quickly. By 1990 revenues reached 1 billion dollars, with more than 10,000 pay-per-call programs available.

Although the growth of the 900 industry was nothing short of explosive in its formative years, it has experienced some definite growing pains along the way. Because the quick profit potential is so attractive, many less-than-reputable players jumped onto the bandwagon early with easily accessible (by minors) dial-a-porn programs or straight rip-off programs. Predictably, with virtually no rules or regulations in place, there were many abuses, and the 900 industry earned itself a bad reputation which it is still trying to shake off.

As you may already know, dial-a-porn programs haven't been the worst examples of sleaze in this industry. Indeed, most of them delivered exactly what was promised, which was why they were so wildly successful. The real abuses have been with dishonest variations of the sweepstakes, credit card and job search lines. While a handful of these programs were legitimate, the vast majority were unequivocal rip-offs of the worst kind, preying upon unfortunate people who can least afford to part with their hard-earned money.

There was little regulation in this emerging industry, and a lot of unscrupulous people took advantage of this fact. Many people still equate 900 with sleaze. This is now changing, and the industry is working hard to improve its image. Federal laws and regulations are now in place, helping to clean up the industry by placing clear, unambiguous limits on what can and cannot be done over a 900 telephone line. More and more reputable companies are using 900, resulting in hundreds of legitimate, useful, and valuable 900 information services, as you will read about in this book.

There are plenty of opportunities in valuable, helpful information or entertainment applications. Like any maturing industry, the questionable fast-buck days are over. You can no longer count on the novelty of 900 to make just about any 900 program successful. In order to successfully compete now and into the future, you must offer sound value for the money you charge. This holds true for any product or service, and the 900 industry is no exception.

AT&T recently commissioned a strategic study on consumer attitudes to 900 services, conducted by the Monitor Company of Cambridge, Massachusetts during the first nine months of 1994, interviewing over 1,100 consumers who use the telephone at least four times per month for information calls to businesses and organizations. The results were surprising. Two of the biggest barriers to calling 900 numbers were uncertainty about cost and lack of interesting programming, with cost uncertainty being the most serious barrier. This may indicate a flat-fee charge would be more acceptable than per-minute charges, which is certainly possible with programs of a predictable duration.

Although the Monitor study indicated that consumers were willing to use 900 numbers, a large number (64%) indicated that a lack of interesting programming kept them from calling. Other barriers were too expensive (40%), negative perception (40%), ripoff (36%), and blocked access (7%).

The study further determined that consumers were willing to use 900 numbers for specific program categories including professional services such as medical information, financial, auto repair and premium customer services like computer support. Also identified as desirable were government services such as tax assessment, professional licensing verification, background checks for firearm sales and motor vehicle applications.

The Future of 900

What accounts for the explosive growth of the 900 industry? Or, by extension, the business of selling information by telephone? Is it a fad that will fade away after the novelty wears off? Will people continue to pick up the telephone, knowing that they are paying for the information? Will the industry still be around a few years from now?

First of all, we are already in the habit of using the telephone for getting information quickly. We call the airline for flight information, we call our stock broker for the latest price of Disney stock, we call the IRS help line for tax questions, or we call the weather service for local forecasts. We pull out the *Yellow Pages* and let our fingers do the walking. The telephone is the quickest and easiest way to get specific information exactly when we need it.

Why in the world would we pay for information when it can be had for free? Using the above examples, the airline may put us on hold for several minutes, we may trade phone calls with our broker a few times before speaking to him, we might get inept tax advice from the IRS (not unusual!), or we may want more specialized weather information.

Not being able to get exactly the information you want when you need it can be very inconvenient in this hectic electronic age. Will we pay for accurate, timely information? Absolutely! As long as the cost of the information seems to be reasonable when balanced against the extra convenience of getting what we need when we want it. We are in love with the telephone because it offers convenience, availability, accuracy and anonymity.

Definitions

Before reading any further, you need to understand several commonly used terms in the 900 industry. This is by no means a comprehensive glossary, which can be found at the end of this book.

Audiotext (also Audiotex). This term broadly describes various telecommunications equipment and services that enable users to send or receive information by interacting with a voice processing system via a telephone connection, using audio input. Voice mail, interactive 800 or 900 programs, and telephone banking transactions are examples of applications that fall under this generic category.

Information Provider (IP). A business or individual who delivers information or entertainment services to end users (callers) with the use of communications equipment and computer facilities. The call handling equipment is often not owned by the IP, and a separate service bureau is hired for this purpose.

Interactive. An audiotext capability that allows the caller to select options from a menu of programmed choices in order to control the flow of information. As the term implies, the caller truly interacts with the computer, following the program instructions and selecting the information he or she wishes to receive.

Interexchange Carrier (IXC). This term technically applies to carriers that provide telephone service between LATAs (see below). Long distance companies such as AT&T, Sprint, and MCI are also known as interexchange carriers.

Local Access Transport Area (LATA). This is a geographic service area that generally conforms to standard metropolitan and statistical areas (SMSAs), and some 200 were created with the breakup of AT&T. The local telephone companies provide service within each LATA (Intra-LATA), while a long distance carrier (IXC) must be used for service between LATAs (Inter-LATA).

Local Exchange Carrier (LEC). This is the local telephone company that provides service within each LATA. Also included in this category are independent LECs such as General Telephone (GTE). The LEC handles all billing and collections within its LATA, often including long distance charges (Inter-LATA), which are collected and forwarded to the appropriate interexchange carriers.

Pay-Per-Call. We have already talked about this with regard to 900. The callers pays a predetermined charge for accessing information services. This is not, however, the only type of pay-per-call service available. For local, intra-LATA applications, a seven digit number is available with a 976 or 540 prefix. This service is usually quite a bit less expensive than long distance 900 services, and should be seriously considered for any local or regional pay-per-call applications that will not have the potential for expanding nationwide.

Pay-per-call services may also be offered over 800 or regular toll lines using credit card or other third party billing mechanisms. When the caller pays a premium above the regular transport charges for the information content of the program, regardless of how payment is made, it is considered a pay-per-call service (the FCC's definition of pay-per-call, however, includes only 900 numbers).

Regional Bell Operating Company (RBOC). These are the seven holding companies that were created by the breakup of AT&T (also known as Baby Bells):

1. NYNEX
2. Bell Atlantic
3. AMERITECH
4. Bell South
5. Southwestern Bell Corp.
6. U.S. West
7. Pacific Telesis

These companies own many of the various LECs. For example, NYNEX owns both New England Telephone and New York Telephone. However, there are numerous independent LECs that are not owned by any RBOC. For example, Southern New England Telecommunications Corp. (SNET) is an independent LEC serving most of Connecticut's residential customers, and has nothing to do with NYNEX.

Although not a RBOC, a major telephone company that should be mentioned here is GTE. GTE is the largest US-based local telephone utility providing voice, data and video products and services through more than 22 million access lines in portions of the United States, Canada, South America, the Caribbean and the Pacific. GTE's intra-LATA 900 service is currently available in California, Florida and Indiana, and should be investigated by anyone wishing to do business in those states.

Service Bureau. A company that provides voice processing / call handling / audiotext equipment and services and connection to telephone network facilities. For a fee, these companies allow an information provider (IP) to offer a pay-per-call program using the service bureau's equipment, expertise and facilities.

What Makes a Successful Program?

A savvy 900 veteran said,"Most 900 programs fail because they are underfunded and underplanned." The elements that make up a successful 900 number information service are the same as those found in any other successful enterprise:

1. A real demand/need for the service. This may sound a bit obvious, but many programs have been tried and failed simply because there was never any demonstrated need for the service. Like trying to sell ice to Eskimos in January. The most successful programs have invariably appealed to a fundamentally important human need, usually related to health, wealth and love/sex. It goes without saying, the market must not only exist, it must also be large enough to generate sufficient call volume.

2. Sufficient capitalization. Most businesses -- not just 900 number businesses -- fail because they don't have enough money to hang in there long enough to turn the corner and succeed. This business in particular is characterized by a long lead time between getting the phone call and seeing the money for that call.

3. Sufficient and appropriate marketing. Location (location, location) is to real estate what marketing is to 900: by far the most important criteria leading to financial success. This is a direct response advertising medium, which takes a lot of skill, savvy and money in order to succeed. Although most failures in this area result

from lack of capital, few beginning entrepreneurs take full advantage of the free publicity and other guerrilla marketing strategies available out there. The marketing program needs to be carefully planned, highly imaginative, and relentlessly aggressive.

4. Take full advantage of the medium. This medium is best suited to information that is time-sensitive or highly specialized. The most successful programs incorporate both types of information.

5. Real value, delivered at a fair price. The result of delivering a quality service that meets or exceeds expectations for a fair price is CUSTOMER SATISFACTION. This reduces chargebacks, creates word-of-mouth advertising, and earns repeat callers.

As you can see, these elements will apply to any successful business. There is no secret as to what makes a winning program. Remember, 900 is not a business in and of itself. It is nothing more than an information delivery medium. The IP must always remember that his business is selling information, and that the telephone is simply one of his options for reaching his customers.

Again, the purpose of this book is to give you ideas. By getting a feel for what's out there, what's been tried and what's still around, you will develop your own perceptions about what programs might work. It may be a completely novel program idea or an imaginative variation on an existing program.

If you don't get enough ideas from reading these profiles, go to the reference section of your local library and see what a wide selection of information is offered through hundreds of reference volumes covering every conceivable information classification. This information is obviously valuable to quite a few people or reference volumes would not have been published to cover it in the first place.

The future will belong to those imaginative infopreneurs who identify a specific information need for a well-defined target market, then who design a responsive program that delivers critically important information in a manner that is superior to all the alternative delivery methods, taking full advantage of the capabilities of interactive audiotex services.

Chapter 1

Chapter 2
Applications

There are two ways of classifying telephone-based information programs: by the method of delivery or by the information content. The remaining chapters in this book are divided into general categories relating to the information content of the programs. The actual method of delivery is clearly of secondary importance to the purpose and content of the program.

Nonetheless, before getting into the profiles of the scores of information programs out there, we must first understand the different telephone delivery options:

Passive. The caller simply listens to a recorded message of a specific nature and duration. Many of the early 900 programs were passive in nature, however few programs continue to be passive due to the limited information that can be offered. However, many polling applications are passive, where the call is simply tallied as a vote by calling the telephone number that corresponds to the vote to be cast.

Interactive. This option is generally available only to callers with a touch-tone telephone, and usually consists of recorded information that is categorized in some logical fashion. The caller is given a menu of selections, to be chosen by pushing the appropriate number on the telephone keypad. There can be numerous sub-menus within the main menu, so that the caller can quickly zero-in on specific recorded information. For a game application such as a trivia quiz, the keypad is used for answering multiple choice questions. In some applications, the caller may be able to leave verbal information, such as name and address, when ordering merchandise or entering a contest. Interactive is by far the most prevalent 900 delivery option, and the majority of the 900 applications we will be covering fall into this category.

Live Operator. There are some applications where recorded information is inappropriate, such as legal, financial, medical, or customer service advice. Because it is expensive to hire such professionals as operators, the charges for these calls are typically higher than the recorded passive or interactive services.

Facsimile (FAX). This is a promising new interactive application whereby a caller with a FAX machine can receive a hard copy of the desired information. This service is also known as fax-on-demand or fax-back. Any information with long term-value, or that is difficult to convey verbally -- such as charts, graphs, or detailed financial reports -- is a good candidate for an interactive FAX program. Business-to-business services are the most logical fax applications because few households are as yet equipped with fax machines.

Computer/Data. There is no reason that 900 numbers cannot be used for access to computer bulletin boards or on-line services. Although few service providers, if any, are using 900 numbers for this purpose as of yet, this application is ready for some pioneers.

Hybrid. Many interactive programs have a "default" option for rotary telephone callers that results in a passive call where the caller hears the main message. Or, an interactive program may offer a live operator on the menu of choices should the caller be unable to get the information he or she needs on any of the recorded menu options.

Timely Information

What types of services and applications are suitable to the telephone delivery medium? Timely, or real-time, information is probably the most obvious application. Any information that changes quickly or continuously falls into this category. Instant access to this information is very useful to the caller. Stock market quotations, foreign currency exchange rates, commodities prices, sporting results, and weather forecasts are some examples of timely information. All of this information will eventually be available later in print or on television, but many people need this information quickly, and will pay a reasonable fee for instant access.

Specialized Information

Another type of information that is suitable for 900-number telephone applications relates to specialized information which, although available elsewhere in publications or from experts in the field, can be better delivered through the use of a 900 telephone service. The telephone may be more convenient, permitting the user to call at his or her convenience when he or she needs the information, instead of waiting for an appointment or driving to the library. A 900 number can be significantly less expensive, allowing the caller to get very specific legal advice, for example, without having to schedule a minimum one hour appointment with a $100-per-hour attorney. A 900 number offers the caller complete confidentiality and anonymity, so the caller can avoid the embarrassment of talking about his or her substance abuse problem face-to-face with the counselor.

Other logical examples are medical advice, customer service assistance, income tax preparation, movie reviews, or even Tarot Card readings.

Customer service help is one specific application that is rapidly gaining consumer acceptance. We all hate to be put on hold indefinitely, play telephone tag, or get shunted around to various people before getting our question answered or our problem addressed. We're too busy, and life is already stressful enough without such barriers and inconveniences. Customer support is one of the fastest-growing 900 applications, particularly the computer software support lines. According to Kathryn Sullivan, AT&T's marketing vice president for business applications and information services, "Computer software companies have been among the pioneers in the innovative use of 900 service to improve customer support and satisfaction with pay-per-call technical support." Among the companies using AT&T's MultiQuest 900 service are industry heavyweights Microsoft Corporation and Lotus Development Corp.

As indicated earlier, the following chapters are classified by information content. It wasn't easy coming up with categories that would cover most of the types of information programs out there.

Many of these programs will fit into more than one classification. In these cases, we tried to put the program into the category that was most suitable. We probably didn't always succeed. In the end, we still had to include a "miscellaneous" chapter to cover those programs that just wouldn't fit anywhere else. Nonetheless, what follows is a fairly logical division of information classifications.

Where to go from here

You are probably reading this book because you're looking for ideas for launching your own 900 number information service. You will get plenty of good ideas here. But then what?

You will need to take the next step and learn more about the mechanics of starting up such a business. The resources listed in the appendices will be a big help. Of course, by far the best resources are those written by the authors of this book.

900 KNOW-HOW: How to Succeed With Your Own 900 Number Business by Robert Mastin is widely recognized as the bible for this industry, giving step-by-step advice on how to launch and then prosper with a 900-number business. The perfect companion book to this one.

Audiotex News by Carol Morse Ginsburg is not only the basis for this book, it is the only trade periodical devoted exclusively to the 900-number audiotex industry. Published continuously since 1988, the profiles featured in this book are only a small sampling of the information that is published every month in this comprehensive newsletter. A subscription is essential for any serious player in this industry.

There are many other resources listed in the appendices that will help with your research and education in preparing to enter this business. Indeed, you will find the most complete listing of relevant resources published anywhere.

Chapter 3
Customer Service
& Helping Consumers

As mentioned earlier, customer service is an area that is growing rapidly as consumer acceptance to 900 pay-per-call increases. People realize that the level of service improves when a reasonable fee is paid because the service provider can afford to staff the lines with competent, well-trained personnel. Much preferable to being on hold for an eternity, or playing telephone tag all day.

Also included in this chapter are services that help consumers or customers in other ways, such as pricing guides, product information or dealer/store locators.

Gateway Sets up Tech Support
New Service for Software Developers

A new service, developed by Gateway Telecommunications, can convert a new or existing hotline to a 900 call. Called Tech Support, it enables companies to provide product support and technical troubleshooting using pay-per-call.

Previously, for the most part, the companies using 900 for technical support AN had spoken with were quite pleased with the results. 900 helps cut down on the average length of time a call takes, because callers are better prepared when they are paying per-minute charges. This frees up more technicians to answer questions -- cutting down on the number of busy signals. In some instances, the revenue generated has enabled the companies to hire more technicians.

A win-win situation for all concerned. "Software developers, electronics manufacturers and high-tech companies whose customers can't afford equipment down-time have been looking for a program like this," stated Arthur Toll, Gateway Telecommunications president. "Basically, we've removed all the administrative headaches from the vitally important hotline function."

Gateway's call diverting abilities automatically switch incoming calls to the client's number, and personal identification number (PIN) generation capability enables this to work in conjunction with existing hotlines. Each incoming call gets a PIN which entitles the caller to a designated amount of hotline usage. *July 1991*

Soft and Hardware 900 Support

Microsoft Corporation, a software manufacturer, charges $2 a minute to connect users of its MS-DOS software to technicians for support services. Some software and hardware companies are prioritizing their customer service with 800 and 900 numbers. Customers with maintenance contracts get an 800 number for service questions; those without contracts get a 900 number to call. Companies can track individual 900 usage, and if they notice heavy usage, can suggest a maintenance contract. Several companies are offering to apply the accumulated 900 fees as a credit toward the maintenance contract. *Oct. 1991*

Worldwide NetWare
900 Line Offers NetWare Support

900 Support Inc. expanded its 24-hour Novell NetWare hotline on Sept. 1 to include support for overseas users. The hotline is staffed by 12 certified NetWare engineers.

When the application began in 1989, it used a pay-per-use 900 number. In April 1991 it added an 800 number so that users at companies with 900-number restrictions on their phones could still call the hotline. Local Area Network users, system administrators, engineers or resellers use this service. Networks by their nature

---

cannot be taken down for maintenance during the business day, so offering technical support 24 hours a day is an obvious advantage.

Computer Reseller News reported that technical help from 900 support is much cheaper than going to Novell directly. The price of the call is $4.99 for the first minute and $2.99 for each additional minute. Callers using the 800 number can either establish an account or charge the service to a credit card.

In addition to the service charge, overseas callers will be billed by local phone companies for the call. In the US, call 900-Pro-Help or 800-Yes-4Tec. Customers outside the U.S. can call 503-Net-Help. *Jan 1992 (active)*

Yellow Pages & Then Some

American Business List of Omaha is marketing a database of Yellow Pages listings -- over the phone. Callers to Business Infoline pay $2.00 for the first minute and a dollar a minute after that. They can get the names and addresses of florists in Reston, Va. or hardware dealers in Evanston, Ill. from a live operator. The number is 900-896-6000, and the caller has access to a listing comprising more than 4,800 Yellow Pages nationwide, with a database of more than 13 million listings. *Feb. 1991*

Want To Buy A Car Or Truck?

National Automobile Data Service, Inc. (900-999-CARS) has introduced a computerized system that makes information on current used auto and truck values accessible. Within an average of four minutes the caller can obtain the current high and low wholesale retail values of models from 1975 to the present. The data is updated and the service includes a list of applicable vehicle options. The cost is $1.75 for the first minute and 75 cents for each additional minute. *June 1990*

Instant Article Retrieval Service

Omaha, Nebraska-based Telemarketing Technology, Inc. has introduced Publishers Instant Article Retrieval Service. This new service allows readers to dial a 900 number and have an article or index forwarded to their fax machine in 30-60 seconds. TTI has also introduced the Executive Reader Response Card Service, whereby readers dialing an 800 number can receive advertising information by fax machine. *Oct. 1990*

Pratt & Whitney High On 900
Heavy-Industry Manufacturers Benefit

An innovative 900 application in West Hartford, Conn. is the first use of 900 for heavy-industry manufacturers. Pratt and Whitney, the 130-year-old machine tool manufacturer with more than 130,000 serialized products in the field, now allows customers to receive technical advice over the phone using 900 technology.

Mr. Edward P. Tenney, general manager for After-Market Operation, enthusiastic about the service, explained the dollar breakdown, "AT&T gets 10 percent of the total billing, plus 35 cents for the first minute and 25 cents for each additional minute, plus the monthly set up costs."

Previously, the 600-odd calls a month the company received on its standard lines were economically unfeasible. At $2 a minute, callers are motivated to be better prepared with the information needed for technicians to analyze problems, so calls are taken care of faster. From a previous average of 15-20 minutes per call, they are down to 7 minutes. Now the service is profit-making.

For Mr. Tenney this is the aspect he is "most delighted with." He added, "Now everybody is handled. Only two problem areas remain to be worked on: Large companies that have blocking and customers from Canada, South America and Europe who cannot use the 900-number line."

As is true for much of industry, the arguments for Pratt and Whitney going 900 were many. Here was a service with a value and while no customers like to pay for something they have been getting at no cost, most understand that free technical support can't go on forever. Equipment had been resold several times and new owners were glad to have the technical support at reasonable prices particularly when the average cost of operating a machine is between $80 and $120 per hour. Additionally, the customer's cost of using the 900 service compares favorably with the expense involved in sending a technician to the customer's location.

This Telephone Diagnostic Assistance service is provided over an AT&T MultiQuest Service 900 Line. Logs of all 900 calls kept by Pratt & Whitney combined with detailed call summary reports from AT&T provide managers with more detailed customer data and allow for more efficient use of technicians. *Aug. 1990*

Help Line for Computers

900-Pro-Help specializes in questions dealing with NetWare and Windows 3.0 computer software programs. About 40 percent of their calls are from resellers who are installing systems, getting into trouble and happy to pay the $2.99 a minute the call costs.

The round-the clock service puts technical support within instant reach of network systems managers and engineers during off-hours when installation configuration and reconfigurations are usually done. *April 1991 (active)*

AAA Traveling the 900 Road
Members Only Get New and Used Car Prices

In what is no doubt an effort to remain the main source of information for its members, the Automobile Association of America (AAA) has joined the *Consumers Union* magazine and the *Blue Book* in providing auto-related information to the public via 900.

The service is available only to members, who can receive the following information:

New cars: Dealer invoice price and manufacturer's suggested retail prices for the vehicle are options. A printout is mailed the next business day after callers receive the phone data.

Used Cars: Dealers retail, private and trade-in prices are available. Pricing data given over the phone is relative to the caller's geographical area. Estimates are available for the past 13 model years.

Members may price up to three vehicles per call. The service costs $1 for the first minute and a $1.95 for each additional minute. The typical calls last six minutes.

AAA's Auto Pricing Service is updated daily, based on prices published by various sectors of the automotive industry. The equipment is capable of responding to 1,000 calls per hour.

Used car pricing is available 9:00 A.M. to 2:00 A. M. seven days a week, and the new car information is available until midnight Monday through Friday, and 9:00 A.M. to 7:00 P.M. on Saturday. *Sept. 1991 (active)*

900 Helps Day Care Search
$19 Call for Parent Help Information

Parents in need of day care in Minnesota's Twin Cities, St. Paul and Minneapolis, can now turn to a 900 number. For $19 per call, Care Connect at 900-680-CARE offers parents in the seven-county metro area a list of licensed child care providers near their homes.

The line, which began September 14, is sponsored by Resources for Child Caring and Consumer Alternatives, two organizations that offer child care referral lists by mail. "We have some high hopes for the line," said Theresa Fjelstad, owner of Consumer Alternatives. "We think we'll be able to attract a large population of parents looking for child care."

Fjelstad said she expects the service to receive 3,000 calls its first year, and grow to have a call count of 25,000. The line should generate some profits by the third year or so, she said. They are using Milwaukee-based Schneider Communications as a service bureau and utilizing AT&T lines.

Fjelstad said that all of the seven counties covered by Care Connect have county child care referral services, but not all parents use them. "According to our studies, 40 percent of parents out there don't use county child care referral," she added.

Marketing efforts for Care Connect will include radio advertising and materials in the packets given out at area hospitals' prenatal classes. According to Fjelstad, a number of the child care providers Care Connect lists have volunteered to spread the word about the line. "The providers themselves are one of our most valuable marketing tools," she said.

Care Connect offers callers a list of child care providers after they have punched in their zip code, child's age group, hours they need care, if they prefer a non-smoker as a care provider, and if they are looking for home or center care. Callers can speak to a consultant during business hours and request that the list of referrals also be mailed to them.

The line carries a disclaimer advising parents that the service cannot vouch for the quality of the child care providers it lists, so parents should do their own screening.

According to Fjelstad, if Care Connect proves successful after six months, it might be offered in other states as a franchise. *Oct. 1992*

Airline Antitrust Line
Works to End Consumer Confusion

In the wake of the $415 million settlement of a class action suit concerning alleged price fixing by four major airlines, Advanced Telecom Services Inc. has created the Airline Antitrust Line. It is designed to end consumer confusion about rebate eligibility.

Estimates show that more than 10 million Americans are entitled to a rebate of some sort from any of the four airlines involved -- American, Delta, United and US Air.

Callers dial 900-884-7767 to hear a four-minute recorded message giving eligibility requirements and how to order rebate forms. The call costs $1.99 per minute.

In response to heavy demand for rebate forms, a second 900 line has been established for requesting the forms. By calling 900-884-7477, consumers can receive by fax the seven pages of documents necessary to request a rebate. The cost is $9.99 per call. *Oct. 1992*

AKC Helps Give Dog Info
900 Goes to the Dogs

The esteemed American Kennel Club (AKC) is now using 900 technology to assist prospective buyers of a pure-bred dog in their search for the perfect pooch.

The recorded line, 900-407-PUPS, gives callers the name and phone number of a referral representative in their area who can help them locate a local breeder. Calls cost 99 cents per minute and average two to three minutes in length.

The line began last September and the service bureau is N.H.-based Innovative Telecom Corp. The AKC's marketing efforts include flyers distributed at dog shows and press releases sent to the media.

The line is going well so far, said Suzanne Lustig, manager of public relations. Lustig noted that with one in three homes in the United States having a dog, there is a definite public interest in pet ownership.

Through a separate customer service number originating in North Carolina, the AKC's Customer Service Department gives callers educational information on acquiring a puppy, and a geographical listing of dog clubs. *Mar. 1993 (active)*

EAA Uses 900 for Convention Info

A 900 line used this summer by the Experimental Aircraft Association (EAA) is a good example of niche marketing. Callers to 900-884-8485 could hear information on the highlights, dates, admission fees and housing for the 1992 EAA Fly-In Convention.

According to convention chairman Tom Poberezny, prior to its convention, the EAA receives a tremendous amount of calls. The hotline provided the most often requested information. Each call costs 80 cents per minute. *Sept. 1992*

Artisoft 900 Technical Support

The computer network products company, Artisoft, Inc., has turned to 900 to offer customers an option for immediate technical support. At $2.50 per minute, 900-555-8324 is optional and targeted to users of Artisoft's network operating system who need immediate assistance.

Joe Stunkard, public relations manager, stressed that the Tucson-based company does not regard this first use of 900 as a revenue generator, and would still offer "free" unlimited technical support.

Stunkard noted that the company's regular toll number for technical support receives approximately 20,000 calls per month, and callers often have to wait to speak to a technician. The 900 line, launched July 26, receives between 50 to 75 calls per day, "applause" and no complaints.

Artisoft has a 65-member technical staff to handle its technical support calls. The 900 line operates from 7 a.m. to 5 p.m. Mountain Standard Time. Artisoft has contracted directly with AT&T for the line's operation.

According to Stunkard, Artisoft has not aggressively marketed the 900 line. Instead the company publicizes it in a quarterly newsletter reaching 150,000 resellers and end users, and in a monthly newsletter reaching 10,000-20,000 resellers. They also sent press releases to the general and trade press. *Nov. 1993 (active)*

B. F. D Software Support

AN has often reported that many software publishers and video games manufacturers are charging for customer support. According to Bruce F. Dyer, president of B.F.D. Productions, a Las Vegas based service bureau, "The best way to do this seems to be via a 900 number."

"We've come up with a 900 help line package that combines an extensive prerecorded menu system with interactive fax and live support. Plus, the client receives a complete call detail report that tracks the areas of the software which seem to give the end-user the most problems. This helps the publisher know what to change for the next upgrade release."

Some software manufacturers find that the 900 technology helps them regain losses from unauthorized disc piracy. Share-warers who lack the printed documentation usually have the most questions.

B. F. D. reports some of their clients currently using 900 for customer support include: Electronic Arts, LucasFilm Games, Konami, S.S.I., Virgin Games, Activision and Computer Solutions. The prices range from $.70 to $2 a minute, and in some cases more. *July, 1992*

Car Repair Helpline
Consumers Avoid Auto Repair Rip Offs

Some recent calls to the AN office have referred to the start-up of live car advice lines. One that began last month, after considerable study and testing, was the CAR REPAIR HELPLINE -- 900-776-1CAR.

Jeff Parness, the president of American Consumers' Network (ACN), says that the IP sees the difficulties Californians are currently having with Sears alleged abuses of car owners as highlighting the need for this type of line. According to Parness, "The line was designed solely as a consumer protection service."

The line is a joint venture with Tribune Media Services, a media syndication company. The advertisements for the ads appear in New York City, Orlando, Fla., Syracuse, N.Y. and Columbia, Mo. newspapers. Parness anticipates rollouts in other large and small papers across the nation. He said, "Coincidentally, California was next on our list." [No doubt their public relations agency, Litzky Moore, will be using the Calif./Sears issue in press releases.]

Parness told AN, "We were fortunate to tap into a pool of master mechanics who are happy to talk car repairs." Presently ACN has 19 on call with seven more available. The mechanics are all in their own sites and ACN's service bureau, Phone Programs USA Inc. designed the line with call forwarding features and data gathering additions that assist in call analysis and data for ACN to use in fine tuning the program.

The line operates Monday through Friday from 6 PM to 10 PM and from 9 AM to 6 PM on weekends. The first minute of the calls is free; each additional minute costs $2.50. Based on test calls, ACN anticipates the average call lasting four minutes.

The hotline helps car owners communicate better with their mechanics by providing callers with easy -- and immediate -- access to expert and objective "second opinions," crafted to help them avoid potential repair ripoffs.

The team of car care specialists provides: a diagnostic analysis of what might be wrong with the car and a list of specific questions the caller should ask to insure that the car gets fixed right ... the first time.
July 1992

Dial-a-Mechanic Line

Callers in need of a quick opinion on what might be wrong with their cars can dial 900-AUTOHELP. Chicago-based Consolidated Service Corp. has 54 years of experience providing fleet maintenance for a variety of large companies. Pat Betz, director of special programs, said the 900 line was a natural outgrowth of the company's business. Additionally, working with a motor club, they had consumer expertise: checking estimates for repairs, maintenance and pricing.

For $2.95 per minute, callers can speak to an ASE certified mechanic. Operational since July, the Dial-a-Mechanic line operates between the hours of 7:30 a.m. to 6:30 p.m. central time Monday through Friday. Stargate ICI is the line's service bureau.

Betz said they were surprised to get so many calls from men. She speculated that the anonymity of the telephone allows callers to ask questions and not feel stupid.

Consolidated Service Corp. has not yet done any advertising for the line, but used a small public relations firm, Lekas and Levine for the initial promotional efforts. They intend to hire an advertising agency, and air a TV commercial, thinking this will give them the data to assess the line's popularity. *Nov. 1993 (active)*

Health Care Costs

Some of the 900 lines with the longest shelf life are the health lines. The Health Care Cost Hotline, carefully launched in September, shows signs of being no exception.

The line, at 900-225-2500, allows callers to access information on the average cost of particular medical or dental procedures in their area. The line lists procedures by health care service codes previously available mainly to health care providers, insurance companies and benefit administrators.

Through a separate 800 number, 800-383-3434, callers can request a copy of The Hotline Code Book listing those codes. A caller can also dial the 900 number and order the code book through voice messaging. The caller is billed for that 900 call, but receives credit for a later free call.

The line provides cost information for more than 7,000 procedures for every geozip (the first three digits of a zip code) in the United States. It lists a median fee, which is the exact mid-point of fees charged by health care providers, and a meaningful range of fees above and below that. "Prepared with the median fee and a range of fees for a medical or dental procedure in their area, callers can use this information to negotiate with their health care provider or seek more favorable fees elsewhere," said Sandy Taube, spokesman for The Health Care Cost Hotline.

The line is a joint venture between the New York-based consultants Milliman & Robertson (M&R), and Medicode, the nation's largest supplier of health care data and management tools. According to Taube, it was in development for more than four months.

Calls costs $3.95 for the first minute and $1.95 for each additional minute. The average call takes two minutes.

With the help of Laura Rzasa of Donely Communications, the line has already been successful in getting national press. The September 27 issue of *U.S. News & World Report* ran a short article on the line in its "New You Can Use" section.

Coopers & Lybrand's (C&L) IVR service bureau handles the line. Taube said that C&L was selected due to their call capacity. "We have a call center of our own where we operate various employee benefits-based transactional IVR systems. Our system is called MARV (M&R Voice). Our MARV system will be employed to deliver client sponsored 800 Hotline service."

Taube told AN that the database can be customized and marketed to corporate clients for employees' 800 use. He explained that Susan Berkley was chosen to be the voice. "She's an experienced talent whose voice was selected after extensive focus group-based research.

"The quality of the deliverable was of great concern in the development of the service," Taube said. "We wanted to be sure callers received value for their money. Additionally, we were concerned with the quality of the voice, assembly of text messages (seamless as possible), and speech quality.

"A charge under $6 is a small expense when it can potentially save hundreds of dollars," Taube said. "Consumers are becoming frontline soldiers in the battle against rising medical care costs, and The Health Care Cost Hotline's fee information enables everyone to become better informed health care consumers." *Dec. 1993*

Home Buying via 900

Home buyers can make a 900 call part of their search. Drawing from a national database, For Sale By Owner gives callers information on for sale by owner homes in the area code of their choice.

Calls to 900-535-FSBO cost $1.50 per minute. The line's parent company is Property Network of America. Smaller companies contract to represent specific area codes. The nationwide system, which currently includes 116 area codes, is tied into the one 900 line. One of the area code representatives (ACR) is Lyco, Inc., which represents three of the nine area codes in Texas. "For Sale By Owner is an information service, not a real estate service with commissions involved," stressed Bob Collett, vice president.

A 1992 test of the For Sale By Owner system, conducted in Denver, Colorado, generated more than 400 calls per day with an average call length of about five minutes. The 900 line is doing well for Lyco's area codes, Collett said. The San Antonio-based company advertises the 900 line through local newspapers and radio. There are at least 40 other companies also acting as ACRs.

Homeowners list their homes by calling the 900 number and using their telephone's keypad to enter information on their home. Buyers punch in an area code and can talk to the owner selling the home. Home buyers can also receive information on related services, such as real estate lawyers or home inspectors. *April, 1993*

Hot Line for Antique Show

The world of antique collectibles is once again blazing trails on 900. In a market which also has Antique Talking Classifieds *(AN* June 1992), the Antique Show Hot Line has started up. It provides information on more than 200 antique shows and flea markets across the United States.

Callers to 900-903-SHOW can pinpoint a region of the country -- Northeast, Southeast, Western and Central -- and then zero in on a state within that region. Calls to the line, started last spring, cost $1.49 a minute, and the average call is 3 to 4 minutes. The recorded listings give show and flea market dates, locations and phone numbers for further information.

The line has a chance in a market rich in advertising possibilities, given the number of publications specializing in antiques and collectibles, and consumers already accustomed to hunting through classified advertising while pursuing their business or hobby.

The magazine Collector's Showcase offers the other 900 venture for antiques, Antique Talking Classifieds. The line lists descriptions of antiques for sale. *Aug. 1992*

Prison Hotline

Prison News Network, Inc. (PNN) has launched a 900 hotline -- 900-884-CELL. Prison family members and crime victims can locate legal, financial and psychological support.

Calls are $1.49 per minute and possible from a rotary dial phone. Callers' names and addresses are confidential. When the line began in August, its first 45 days were devoted to letting callers tell their stories to learn how PNN could be of help to them.

In addition to helping obtain legal, financial and psychological support, the line is designed to organize prison family members, crime victims and support groups to lobby to influence public policy on education, social welfare and prison reform. It will also carry reports about prisons and crime news stories that the traditional media may not carry. *Nov. 1992*

Product Service Center
MaxServ Inc. Uses 900

Consumers faced with home appliance problems can call a 900 number for help. When *AN* spoke with the MaxServ Inc. staff it was not the uniqueness of this two-year old application that was newsworthy, but the way the number is marketed.

MaxServ has an agreement with Sears, which refers callers to the line when they call the store for advice, or when they purchase parts for installation at home.

Customers are advised to call the 900-84-Sears number when a product is no longer covered by warranty, plus the number is featured in the owner's manual.

Austin, Texas-based MaxServ specializes in repair information systems and services for a broad range of household products, many makes and models.

Technical specialists give do-it-yourself and care and use advice to callers. The call center employs 200 in the Austin headquarters and 35 in the recently opened Scottsdale, Ariz. branch.

The company uses Sprint Telemedia as its service bureau. Bob Maddock, the general manager in Scottsdale, says the company receives 10,000-12,000 calls a month, with the average call in the three to five minute range. The cost is $1 for the first minute and $2 for each additional minute.

Mike Wenig, vice-president for marketing, sees "the match-up of 900 and the consumer trend toward do-it-yourself as having significant potential." *Feb. 1993 (active)*

Shopping Hotline

Avid shoppers in Southern California can now dial the Fashion Addicts Hotline, a 900 number offering information on sales and available items in area stores. Each call to 900-289-0918 costs $2 per minute and runs a maximum of 10 minutes.

Information provider (IP) Helaine Bruck Ross, a costumer for the entertainment industry, started the line. According to Ross, the line is ideal for tourists and can help area residents in their shopping efforts. Some of the stores listed offer discounts if shoppers mention the hotline. *Aug. 1993*

Times Uses 900 Fax

The *New York Times* is now offering their readers a 900 number for use in requesting faxes of computer-related articles. Callers to 900-737-4446 pay $3.95 to have the article of their choice faxed to them. An 800 number allows callers to charge the $3.95 per article to a credit card and order more than one article at a time. For a list by fax of available articles, call 800-551-0159. *April, 1993*

Toy Safety Tips Through 900

Parents and grandparents baffled about the most appropriate and safest toys can find help through Toy Tips Inc. at 900-420-3714 ext. 308. This line identifies the newest, safest, most educational products in the vast toy market.

The line began in 1991 and is generating good minutes, said IP Marianne Szymanski. It offers a different message each week and costs $2 per minute.

Szymanski noted that most callers to Toy Tips stay on the line for the entire five-minute message. She was reluctant to tell Audiotex News who her service bureau is because she is unhappy with them. It can take up to two weeks for this service bureau to return phone calls, Szymanski said.

Szymanski markets Toy Tips mainly through public relations and personal appearances. Further promotion is through a quarterly magazine, The Toy Tips Safety and Selection Manual. (800-929-Toys) -- $20 for four issues. Capitalizing on the nature of the line, she can also generate articles in the general press.

Szymanski's desire for speed and efficiency led her to use 900. "I have so much information, there's no other way I could get it out every week," she said. Szymanski tests each toy she suggests with the real expert -- kids! She has a toy testing team of 130 children at Marquette University's Krueger Child Care Center in Wisconsin. *Mar. 1993 (active)*

Used Car Info Three Years & Line Still Going Strong

Consumers interested in buying or trading in a used car can find important information through the *Consumer Reports* Used Car Price Service, a 900 line going into its third year of operation. The average call to 900-446-0500 takes five or six minutes and gives a particular car's market value.

Consumer Reports developed the $1.75 per minute line to give consumers easy access to that info, said Thomas Blum, manager, auto price services. "If you know the market value, you can bargain for the best price."

Although he declined to give actual call counts, Blum said that the Used Car Price Service is doing well, and *Consumer Reports* is satisfied. The line also brings a high rate of caller satisfaction, Blum said. According to Blum, the 900 line fits in well with the other car-related information *Consumer Reports* distributes. "That's why it works." The magazine has regular reports on cars, as well as a yearly car issue. It also lists cars in its annual Buyer's Guide.

The line is marketed primarily through advertising in the magazine and lists market values for domestic and foreign cars, trucks and vans from 1984 to 1992 models. According to Blum, the figure given takes regional differences into account, as well as a car's mileage, major options and condition. The line operates between 7 a.m. and 2 a.m. Eastern Time. *Jan. 1993 (active)*

Wood Floors Hotline

The National Wood Flooring Association has a 900 line for homeowners' questions on installing and maintaining wood floors. Calls to 900-646-WOOD cost 90 cents per minute. The line operates from 8 a.m. to 5 p.m. Mon.-Friday. *Oct. 1993 (active)*

WordPerfect - 900 - Tech Support

The use of 900 in computer software technical support continues to grow. Last month WordPerfect Corporation introduced its Priority Service option for its business applications, including round-the-clock support for the word processor program, WordPerfect.

Customers using Priority Service can choose either 800 lines with a flat fee of $25 payable by credit card, or 900 lines charging $2 per minute. This is WordPerfect's first use of 900 or 800 credit card billing for technical support, noted Lyle Ball, public relations specialist. "The creation of the program was very customer driven," Ball added.

According to Ball, during calls to request technical support, many customers have expressed willingness to pay for support calls if they could get assistance quickly, 24 hours a day and from the best technicians. Ball noted that the company averages 11,000 - 12,000 completed calls for technical support per day.

Priority Service runs on AT&T lines for both the 900 and 800 aspects. The 900 lines use AT&T's MultiQuest® and the 555 exchange for business applications. Ball said that after a full evaluation of the other long distance carriers, WordPerfect found that AT&T's service best met its needs.

Approximately 900 technicians handle the Priority Service calls from 7 a.m. to 6 p.m. MST, Monday through Friday. The lines for the Windows and DOS versions of the WordPerfect word processor program are open 24 hours a day, seven days a week, excluding holidays.

WordPerfect is publicizing Priority Service through notices packaged with its software products and press releases sent to the media. In addition to the April 4 launch of Priority Service for business applications, Priority Service for workgroup applications and WordPerfect's product DataPerfect began March 1. The company continues to offer free support to customers registered in its Classic Service Program. *May 1994 (active)*

Internet Helpline

Computer-related applications are no strangers to the 900 industry, and last month a 900 number began that is designed to give callers information on the vast computer network, the Internet. The 900 Internet Helpline, sponsored by the monthly magazine *Internet World*, offers callers messages in five information categories.

The line can be reached through either 900-45INTERNET or 900-555-INET. Calls cost 95 cents per minute with a maximum call length of 15 minutes.

A major public relations campaign by Edelman Public Relations accompanied the line's September 14 launch. The campaign covers major newspapers, magazines and television stations, said Alan Meckler, chairman and CEO of *Internet World* publisher Meckler Media. A major advertising campaign targets the computer sections of major metropolitan newspapers, starting with the *New York Times*.

A 900 line complements Meckler Media's other work, including the magazine itself and its twice yearly trade shows. Meckler said, "Our goal is to be the leading source of Internet information for consumers and professionals." As far as Meckler can ascertain, the line is the first to cover the Internet. B.F.D. Productions in Las Vegas is the line's service bureau.

According to Meckler, by November his company plans to expand the line by adding the option of speaking to a live consultant for $3.95 per minute. *Oct. 1994*

Canadian Papers Refuse Auto Ads

Five Canadian newspapers have turned down part or all of advertising from an automobile consumer hotline; two expressed concern over possible danger to auto industry ad revenue. For $15 plus tax, Green Lite Information Corp. gives callers invoice costs for new cars and average trade-in values for used ones.

The Toronto Sun, The Toronto Star, Branford Expositor and *The Hamilton Spectator* refused to run the ad. *The Kitchener-Waterloo Record* only runs advertising of Green Lite's used car pricing service. In a letter to Green Lite, and ad supervisor at *The Hamilton Spectator* cited the positive relationship the newspaper has with dealers and national manufacturers and expressed concern that this could be jeopardized by taking Green Lite's ad. A letter from the *Branford Expositor* said the service Green Lite promoted "could be considered conflictive to most automotive advertisers." *Sept. 1994*

In-Depth Profile:
Microsoft® Product Support

Although the Microsoft Corporation has been using 900 for technical support for several years, beginning October 1, 1993, working directly with AT&T, it began using AT&T's Vari-A-Bill and 555 prefix combined with a new pricing structure.

Microsoft is the worldwide leader in software for personal computers, receiving some 23,000 calls a day for technical support. Before developing the Microsoft Support Network, the company listened to hundreds of customers about their preferences in how to access, utilize and pay for support. "We've incorporated customer feedback into all aspects of this support program," said Steve Ballmer, president at Microsoft. He added, "What have customers told us they want in technical support? The right answers right now!"

The 555 prefix allows companies to unblock their phones for 900 calls with that specific prefix, which is reserved by AT&T for business uses only. The idea is that all 900 numbers except those that will help businesses can still remain blocked, preventing employees from running up charges for unauthorized access to 900-number entertainment programs.

The Vari-A-Bill product, introduced in 1992, gives the IP greater flexibility and control. The options range from a free call (can be determined any time during the call); a flat charge (call price is fixed at a set fee from a point the sponsor signals), new rate (changes the per-minute charge), premium charge (flat charge to be applied in addition to the per-minute rate), and premium credit (flat amount to be deducted from the total price of the call).

Microsoft offers a wide range of electronic support services and information available to all of its customers in supporting some 170 different products. Options are available 24 hours a day, seven days a week, many at no cost. In order to simplify its offerings to specific user categories, Microsoft has segmented its products into four clusters:

1. Desktop Applications. Includes Windows and Macintosh applications, such as Microsoft Word, Excel, PowerPoint, Office, Access and FoxPro.

2. Personal Operating Systems. Includes Windows, MS-DOS, Windows for Workgroups, and hardware.

3. Development Tools. Includes Microsoft Visual Basic programming system, FORTRAN and C++.

4. Advanced Systems. Includes Windows NT, Microsoft Mail, SQL Server and LAN Manager.

The desktop applications category is the only one that offers free technical support for an unlimited time to qualified purchasers on regular toll lines (during normal business hours). The other categories offer free technical support for limited periods after purchase only, typically 90 days. Thereafter, there is a charge for any further support. Purchasers of desktop applications can get support after hours on a 900 number. Purchasers of O.E.M. (Original Equipment Manufacturer) computers with bundled Microsoft OEM-licensed software installed are typically not given free support, but can use any of the paid options that are available.

There are several levels of support that can be selected by the customer . For example, one tier of support, called Priority Support, provides the following options:

1. Priority Comprehensive. This provides support across all four product clusters outlined above, with three different purchase options: $7,550/year, $995/10 pack, $150 an incident [900-555-2100 - $150 flat rate call with unlimited time];

2. Priority Development with Desktop. This provides support for everything except Advanced Systems: $1,495/year, $750/10 incidents, $2/min. [900-555-2300];

3. Priority Desktop. This provides support for desktop systems plus personal operating systems: $195/year, $2 per min. [900-555-2000, has a $25 cap because the 800 option is more economical after that] $25/incident [800-936-5700]. There is also an 800 number to guide users through the options - 800-936-3500.

Before this new program was initiated in October 1993, 800-number credit card calls and 900-number calls were about equal in number. Now, calls to the 900 numbers amount to an average of 1,000 per day, while 800-number calls average about 350 per day. This could be an indication that companies are in fact taking advantage of the 555 exchange, having it selectively unblocked, allowing access to more people from their workplaces.

Of course, cost could also account for the difference. All 800-number credit card calls cost the customer a flat fee of $25, regardless of how long the call lasts. On the other hand, 900-number calls are $2 per minute up to a maximum of $25 (Priority Desktop option), which is obviously advantageous for someone with only one or two quick questions.

On the 900 numbers, call charges do not begin until a support engineer comes on the line. The average hold time in queue has been under 60 seconds, which is quite impressive by any standards. The hold time on the regular toll lines is longer, according to Mary Massengale, the marketing manager for product support, but the goal is to cut hold times below 60 seconds on 90% of all product support calls, not just the 900-number calls.

Microsoft has made a serious commitment to customer service by providing such a flexible, responsive program. Further, Product Support is an independent division within Microsoft, and vice president Deborah Willingham reports directly to president Steve Ballmer. Product Support employs some 2,500 people, about 20% of the total employees at the company. The personnel, including service representatives and support engineers, are divided between the headquarters in Redmond, Washington; Colinas, Texas and Charlotte,

N.C. According to Massengale, "Product Support is the fastest-growing group within Microsoft."

When asked whether or not the 900 number side of the product support service was meeting expectations, Massengale was quite positive, indicating that its main advantage is being a per-incident, pay-as-you-go service with the flexibility to fit the customer's specific needs, available 24 hours a day, 7 days a week.

In later years this may be viewed as a milestone for the 900 pay-per-call industry and mark the turning point in public perception and consumer acceptance. Many other software companies are following Microsoft's lead in offering better support services for a fair price, and 900 numbers are becoming quite common as a convenient delivery medium for both the customer and the information provider.

Chapter 4
Government
& Non-Profit Organizations

The use of 900 numbers by governmental and non-profit organizations is one of the most exciting growth areas in this industry. These organizations are always constrained by tight budgets that often prevent them from serving their constituents as well as they would like. A 900 number is often used in order to be able to continue giving out information that was often free in the past, before the days of lean budgets. Basically a telephonic user's fee.

In some instances, non-members/constituents/taxpayers are the ones requesting all the heretofore free information, taking up a lot of staff time in delivering a service to someone who hasn't paid for it. This is not fair to the people who must ultimately pay the salaries of the people providing this information. A 900 number solves this dilemma quite efficiently.

900 Better Business

Better Business Bureaus are considering 900 numbers that consumers would be able to call for assistance. According to Barbara Berger Opotowsky, president of the Better Business Bureau of Metropolitan New York, "things are beginning to move and we are in conversations with service bureaus and telephone companies, for a trial run."

Local bureaus are almost entirely supported by local businesses whose interest it is to encourage honest practices. In many Better Business Bureaus funding is not meeting the current demand for services. "If the cost for a call is one dollar a minute, on an average three-minute call, the BBBs could net $1.80. In the New York bureau, with 350,000 calls last year, that would have yielded $630,000," Ms. Optowsky explained. *May 1990*

Tests For 900
U.S. Chamber of Commerce Surveys

The U.S. Chamber of Commerce and their members have really been giving 900 a test. Working directly with AT&T, it has three applications: an opinion line, a daily updates line on crucial legislation and a business ballot.

The lesson learned from the business ballot for the audiotex industry is that Chamber members can prove to be a truly awesome challenge ... if the call costs.

According to Bruce Josten, a Chamber spokesperson, the business ballot had six questions. The members had a choice of calling at a cost of 50 cents per minute or mailing a postage paid card. In three weeks, eight to nine thousand mailed the card and forty used the phone. *Sept. 1990*

BBB High On Pay-Per-Call
Better Business Bureau Going 900

In May of 1990, when *AN* interviewed Barbara Berger Opotowsky, president of the New York metropolitan Better Business Bureau (BBB), she calculated, "If the cost for a call is one dollar a minute, on an average three-minute call, the BBBs could net $1.80. In the last year [1989] that would have yielded $630,000."

It has taken a year, and Opotowsky had to fight various groups but 900-INFO-BBB is up and running at a cost of 85 cents a minute. The fees will help offset the cost of improving the BBB's overall services in handling and investigating complaints and fielding inquiries.

A free phone line is available for low-income consumers. That number, which was not released, will be distributed through groups that serve the poor. *Aug. 1991 (active, number changed to 1-900-225-5222, at 95 cents per minute)*

Michigan Dept Offers 900

Starting next year, the state of Michigan's Department of Commerce is going to offer 900 service to people and businesses calling for records from its Corporation and Securities Bureau.

The state expects to generate $700,000 from the $1.50 per minute charge for the call. Currently, eight lines and six workers handle the 4,000 calls a week. With hundreds, perhaps thousands, of callers who can't get through, officials anticipate handling twice the number of calls after the adoption of 900 services.

The bureau will hire additional staff for its records and certifications units. With 12 phone lines operating, and additional callers getting through, there will be more requests for certified copies of records. *Nov. 1991*

Buying Lottery Tickets on 900

Massachusetts is considering offering lottery fans the option to buy tickets via a 900 number, and to be billed for them on their telephone bills.

In Indiana, lottery losers are able to call a 900 number, punch in their losing ticket number, and enter a second drawing for a $20,000 prize. *Dec. 1991*

Massachusetts Uses 900
Other States Await Test of Phone Lottery

The Massachusetts State Lottery is now the first state lottery in the country to offer a betting-by-phone service. Baystaters can dial a 900 telephone number and place one bet. The $1.95 charge includes a $1 bet and a 95 cent phone fee. The year-long Mass. experiment limits the calls to within the state.

Bettors dial into the service, key in their Social Security number and choose one of four lottery games. Then they key in their choice of numbers or opt to have the computer pick the number.

The computer is operated at no charge to the state by National Interactive Systems (N.I.S.) of Natick, Mass., confirms the bet with a serial number. The serial number is the "receipt."

Perception Technology Corporation was selected by N.I.S. to provide the voice response equipment. Over a period of time, N.I.S. developed the Lottery-By-Phone Service and implemented it with a Perception Technology voice processing system with a capacity in excess of 1000 lines.

Perception also supplied equipment to the Virginia State Lottery. That system provides an information hotline announcing winning numbers and instructions on how to play various games. It also allows retailers to validate and receive confirmation on winning lottery instant tickets.

Indiana also utilizes 900. There the 900 system currently in place differs from the Massachusetts plan. In Indiana, losers of the instant ticket game call a 900 number at a cost of 60 cents to have the opportunity to win $20,000. In Mass., players have the opportunity to play the big jackpot games.

Although the Massachusetts Lottery is not the largest in the country, it leads in per capita ticket sales -- with $5.30 a week in sales for every person.

Retail ticket agents are concerned that the telephone betting may hurt their sales. Other concerns relate to the potential abuse by children, which is why the bettors were limited to two calls a day and one bet for each call.

Other states are watching this pilot test, and in Maine the lottery Commissioner was approached by several 900 number operators seeking permission to sell tickets over the phone. *Jan. 1992*

Public-TV Helps Viewers Find Videos

Los Angeles public-TV station KCET has a 900 phone line for viewers who are trying to track down a copy of their favorite PBS program. Dick Cook, Director of Distribution, said, "We serve 80 stations and get referrals from those stations."

When viewers calls their local PBS to get a video of a show they have watched, they are referred to 900-860-9301. They began with one operator in September, and now Lo-Ad in Reno, Nevada, acting as their service bureau, routes the calls to four live operators at the station.

"Initially VideoFinders tried to work with AT&T," according to Mr. Cook, "but had difficulties and were referred to Lo-Ad and now are quite pleased with the service they have been getting."

The station put together a database of over 58,000 home videocassette titles. In addition they have developed a listing of over 1,500 public television programs available for sale to viewers. The database is updated quarterly; VideoFinders does high-speed searches by title, genre, performer, etc. for $3.00 ($2.00 to connect with PBS VideoFinders plus $1.00 for the first minute), then $1.00 for each additional minute. *June 1990 (active)*

900 Driving Report Line

A 900 application inspired by winter weather is the Washington State Department of Transportation's Mountain Pass Reports at 900-407-PASS. The line reports on latest weather and road conditions for 35 cents per minute. The service covers precipitation, any accumulation on roadways, temperature, and whether traction devices are advised or required.

Callers unable to access the 900 line can call a toll and 800 number for the information. *Feb. 1994*

900 Catfish

The Catfish Institute of Mississippi, a trade association of catfish farmers and feed manufacturers, has started a 900 line. A 30-second recorded nutritional message is offered, and then a live operator captures the name, address and telephone number of the caller, who receives a cookbook.

The Richards Group is the ad agency that put together the one-minute radio ads that feature chefs and epicures like Lee Bailey, Craig Claiborne, Nathalie Dupree and Wolfgang Puck, who offer their favorite catfish recipes while blues music plays in the background.

Ruppman Marketing Technologies is the service bureau handling the technology and the fulfillment. According to a Ruppman spokesperson, "The line started Feb. 1 and is doing good."

When *AN* spoke to The Catfish Institute, they said they were happy with the $3.95 line because callers were getting their cookbooks in a timely fashion. *May 1992*

Gun Check using 900

Legislation proposing the creation of a telephone 900 hotline number to allow firearms dealers to check the background of prospective gun buyers was recently presented to Illinois Governor Jim Edgar. *Sept. 1991*

Magazine Starts 900 Help Line

Liguori Publications, publishers of *Liguorian* magazine, a well known and highly respected Catholic publication, has begun The Catholic Helpline, a 900-phone line ministry which provides prerecorded information and spiritual guidance. New messages, with all information having ecclesiastical approval, will be added each month. The cost is 83 cents per minute, with each message lasting three to four minutes. West Interactive is the service bureau and has provided the equipment that allows rotary phone owners to access the messages.

Julie Kelermen, associate editor, parish education products, says their main objective is service to readers not making profits. Quality recording was another objective of theirs, and when you try the line, you will see they have succeeded. *Jan. 1992*

National Homily by Phone
Hundreds Call Weekly -- Funds for Church

Lisa Broderick, president of Talisman Communications, a parishioner and director of development for St. Ignatius in Manhattan, has established a nationwide homily by phone service.

According to Broderick, "The service has been running for 9 months and is doing very well. Working with Audio Communications (ACI) in Las Vegas, who have just been terrific, I started the line as a way of demonstrating that 900 could be used, even for small groups, for fundraising efforts."

"The priests load the equipment themselves and everything is working beautifully. The homily is 10 minutes with a flat fee of $10.

With a total universe of 1500 we get hundreds and hundreds of calls each week."

Talisman Communications runs the Pope Line. Talisman is providing its services at no cost to the church, and *AN* is anxious to see their next effort -- it seems New York's Cardinal O'Connor has contacted the company. *May 1991*

Farmers Helped by 900

Because of budget cuts, the Market News Branch of the California Department of Food and Agriculture was faced with discontinuing its free information services for farmers and the agricultural community, until this service was converted into a self-funding pay-per-call program. Now farmers can call 1-900-555-0923 at $3 a minute for the latest farm commodity prices or specialized weather information for better crop management. When faced with either losing access to valuable information that was once free, or paying a reasonable fee, the farmers have opted for paying.

900 Aids DOC Supervision
Offenders Pay Entire Cost of Program

In Delaware, probationers and parolees are now only a 900 phone call away from P/P supervision. This summer, the Delaware Department of Corrections (DOC) started a new telephone reporting system for minimum-risk probationers and parolees.

The system utilizes services and equipment from the Colorado-based BI Inc. and replaces agencies' labor intensive manual systems of caseload management. With more than 2.6 million people nationally on probation, agencies are hard-pressed to keep track of people who are minor offenders.

Through the system, known as BI Profile, each offender dials 900 and a special personal phone number, which serves as their individual reporting code. For its use in Delaware, there is a charge of $4 per call.

During the first week of operation, BI staff enrolled 400 offenders in BI Profile. All must report each month and answer a series of questions about their progress. BI promptly reports any failure to report, or any problem behavior or criminal involvement discovered through phone contacts.

BI runs similar programs in other jurisdictions, including the State of Washington, which has 15,000 clients in the program. The cost to participating agencies is nothing more than start-up and training time. The offender pays for the entire cost of the service.

All offenders assigned to BI Profile call a "900" number and answer a series of questions asked by a computer. The average call takes two to three minutes to complete, and the typical charge is $2.00 per minute.

BI uses AT&T MultiQuest® 900 product and developed its own software. BI does all of the monitoring, recording and archiving at its corporate headquarters in Boulder.

Last year BI acquired TrakTek, Inc. a Phoenix company that was a direct competitor in the electronic home arrest products industry and had marketed systems of its own manufacture and design since 1987. BI's TrakTek system is unique in that it employs 900 Mhz radio technology. Several hardware and software parameters in the Trak II 900 system are adjustable to match the varied supervision requirements of each offender. *Sept. 1992*

Arkansas 900 Aided by New Law

The Arkansas Office of Motor Vehicles joins the growing number of municipalities using 900. Due to go into effect by mid-August, their number is part of the enactment of a new law requiring motor vehicles five years old or newer, with past damage of at least 70 percent of book value, to have "damaged" written on the title.

The 900 number assists car buyers obtaining information on what sort of damage a car sustained and what was done to repair it. The Office of Motor Vehicles will most likely charge a flat $5 fee for the 900 call said Roger Duren, manager of Direct Services. At the time Duren spoke to *AN*, Motor Vehicles was considering contracting directly with AT&T. The live line requires callers to give a car's vehicle identification number.

One planned way to market the line will be on the declaration of damage notice dealers will be required to display on a car's windshield. *July 1993*

BBB 900 Use Is Growing

The 900 usage by regional Better Business Bureaus (*AN*, April '94, July '94) continues to grow. Several bureaus now take consumer inquiries on 900-CALL-BBB. The Better Business Bureau of Maryland, Inc. is one of the latest, going online in mid-October. The Phoenix, Dallas and Houston bureaus also recently went online; but these three bureaus take only out-of-state calls on the 900 line.

Its 900 number has provided the Maryland bureau with a way to effectively handle a large call volume and eliminate the busy signal that callers to the regular phone number often faced. "we consider it huge success already," said David Kershner, bureau president.

Kershner's bureau received 200,000 phone inquiries about businesses in its state last year. Even with eight lines and four operators, busy signals were common, Kershner said.

The 900 system uses both live operators and automated voice response. Callers to 900-CALL-BBB pay 95 cents per minute. Credit card calls are billed a flat rate of $3.80. The 900 system operates on AT&T lines, and Scherer's Communications is the service bureau.

When *Audiotex News* spoke to Kershner, radio advertising to publicize the 900 number was to start shortly, with TV ads scheduled for January. According to Kershner, the latter will be co-op ventures with member companies. The 900 number is not aimed at the bureau's 3300 members, who access a membership hotline.

Dave Major, president of the Arizona BBB, noted that because his bureau only takes out-of-state 900 calls, its traffic would mainly result from referrals given by other Better Business Bureaus. The Phoenix bureau receives up to 1000 calls per day. According to Major, a study the bureau conducted last year showed that almost 25 percent of each day's calls came from out-of-state.

The BBB use of 900 began with the metropolitan New York bureau (*AN*, June '92). Last fall the BBB's national office began a two-year test of 900-CALL-BBB and is moving toward a national network using it. The Philadelphia bureau went online, followed in the next several months by Miami, Buffalo, Chicago, Los Angeles, and Syracuse, plus the metropolitan New York bureau's switch to 900-CALL-BBB. The Seattle bureau is among the next ones scheduled to go online. *Dec. 1994*

Bureaucratic 900
900 Finds a Home With Municipalities

In California, librarians have been enlisted to provide public access to fax capabilities, and would you believe some librarians have been deputized to collect the fees? (Future issues of *AN* will provide additional coverage on some of these services ranging from 50 cents per minute to $5 per call.)

Riverside County makes its weekly Board of Supervisors meeting agenda available to the public via fax.

San Diego Municipal Court offers a service enabling the public to ask a clerk questions regarding their traffic violations.

The Orange County, Calif. Law Library has a line for individuals to access its database with their computer/modems and search for certain documents and court cases.

The U.S. Marshall's Office in Los Angeles offers recorded information about auctions of impounded goods.

The U.S. Bankruptcy Court provides computer screen and audio access to court records for courts in Atlanta, Kansas, Oregon, San Diego and Sacramento.

The City of Oakland is making city permits and forms available, and Ventura County, Calif., is sending out building and zoning information by fax.

The Los Angeles Rapid Transit District provides transit information. *April 1993*

Contra Costa May Use 900

Increasingly, government agencies at the local, state and federal level are using 900 service as a way to disseminate information requested by their constituents.

Faced with hefty budget cuts, officials in Contra Costa County, Calif. are discussing a proposal to start charging for information.

The idea is to charge callers for information that's considered non-essential, or extra. For instance, callers to social services would pay for a recorded message of available counseling services. The probation department might provide tips on improving children's behavior. Probationers might have to pay to call their probation officers.

"... I think the county should look at it as a potential option," said County Clerk Steve Weir, who suggested the 900 lines. County officials are surveying employees to gauge their opinions of the shift. If it goes over well, the Board of Supervisors would consider it.

"We don't want it to be the kind of information the public has a right to expect from us for free," said Claude Van Marter, assistant county administrator. "It really has to be peripheral to a department's operation."

Departments that the public uses heavily, like probation, courts or health services, could make some money. People who call the courts to ask how to file a lawsuit or how to file a small claim, as an example, would pay for those services. *April 1993*

Hunting License Line

The Wyoming Game and Fish Department has joined other municipalities using 900. Its number, 900-884-GAME, allows applicants competing for a limited number of hunting licenses to get the results of the department's drawing.

Results are available immediately with the $5 call. Previously it took up to two weeks to notify applicants by mail. According to Roene Larsen, Game & Fish Department spokesperson, "The 900 line tends to help the department be more customer friendly."

Callers may obtain information for up to three names per call. If the draw has not yet been completed at the time of a call, the caller is notified and hangs up without being charged. *Dec. 1993 (active)*

L.A. Rapid Transit 900 #
Getting Here to There

Call counts are building for the Los Angeles Rapid Transit District's 900 line offering transit information. When the service began in December, only 500 calls were received. The monthly call count is now up to 2,000 -- with word of mouth promotion about the 35 cents-per-minute line.

Doug Anderson, customer information systems coordinator, is looking forward to the call counts when the advertising does start. Anderson said that the live operator line used previously received a great volume of calls with frequent busy signals to callers. According to Anderson, he sold his superiors on the idea of a 900 line by showing the cost deferential of hiring additional staff.

A long-time supporter of 900 technology, Anderson was disappointed at the industry's start and hoped the sex and scams were history. He said consumers, with the average two-minute call, did not have a problem with 900 (Pacific Bell lines) but, as initial surveying indicated, for some users the automated over live information system was a deterrent. Utilizing 900 will off-load a large number of consumers, leaving those who can't cope with the automated system to get information from the live operators, Anderson explained, "This will really improve our service at no cost to the taxpayer." *June 1993*

Major Library System & 900
Computer Users Will Access Data Bases

Computer technology and a 900 number will enhance the work of a major public library system. In a move planned for next year, computer users in southern California will be able to access information from California's Orange County Public Library by dialing a 900 number with their modems.

The library is currently negotiating a contract for system hardware, and installation will start early in 1993, said John Adams, Orange County librarian. Adams expects full implementation by spring. A service bureau has not yet been selected.

The 900 number will be regional, reaching only southern California. The cost will probably be between $1 and $2 per minute, Adams said. The information in the computer system will also be available free of charge to patrons coming to any of the library's 28 branches.

The system will offer data bases, such as indexes of articles in periodicals, the text of those articles, financial and corporate reports, a national telephone directory, bookholdings in libraries across the United States and possibly an encyclopedia.

"The advantages of dial-in service seems so significant, we're sure people would be willing to share some of the costs of the system," Adams said. He noted, because of the ease in billing and collecting, 900 technology was chosen.

When the full marketing effort for the 900 line is underway, some of the first potential users sought will be people who use business information data bases. Other promising market niches are home offices and college and high school students.

The library is optimistic about the line, Adams said. "We're excited to be in the vanguard of applying contemporary technology to the traditional mission of the public library." *Dec. 1992*

Library to Become 900 User

By year's end, Orange County Public Library expects to complete the implementation of a service allowing personal computer users to dial a 900 number with their modems and retrieve information.

The regional line will offer databases and other information at a cost of $1 to $2 per minute *(AN,* Dec. '92). As library officials look ahead to their marketing effort, they were encouraged by a survey cited in the newsletter, Library Hotline, said John Adams, Orange County librarian. The survey said that 85 percent of personal computer owners use public libraries. "It sounds like the market is right there," Adams said. *June 1993 (active, $1 per minute)*

Mich DOC Lines Strong

The two 900 numbers utilized by Michigan's Department of Commerce are generating calls, although the revenue has not reached the heights estimated last year *(AN,* July '92). Still, officials say the revenue generated has been adequate to cover the cost of the lines' operators.

The lines reach the Department of Commerce's Corporation and Securities Bureau and Bureau of Occupational and Professional Regulation, and both charge $1.50 per minute. In mid-'92, the state and AT&T anticipated splitting more than $1 million in revenue, with the state receiving 60 percent.

The state's revenue from the Corporation and Securities line reached $32,000 per month. The line, started January of last year, distributes information on state-registered corporations and limited partnerships. It generates up to 1,200 calls per day.

Throughout last year the Bureau of Occupational and Professional Regulation's 900 line received about 40,000 calls, and brought in some $48,000 through November. The line uses two operators. When it began last February, it resumed a service previously discontinued, verifying licenses of doctors and other professionals. *July 1993*

New Sprint Lines Hotline
Helps Lassie Come Home

Processing calls in Sprint Telemedia's service bureau, the American Humane Association (AHA) has instituted the Lost & Found Pet Hotline. The 24-hour national hotline and tracking system begun last month is designed to reunite dog and cat owners with lost pets. Pet owners can call 900-535-1515 to report lost pets; while those who find stray pets can call 800-755-8111.

In the first-of-its-kind hotline, the voice processing system in Sprint Telemedia's in-house service bureau cross matches information from the 900 line with that from the 800. Through virtually identical menus, callers to each number provide their telephone numbers and location by zip code and the description of the animal they have lost or found by such attributes as type, weight, gender and color. When matches occur, the system immediately voices back the telephone number of the caller who reported finding the pet.

The 900 line costs $1.95 per minute, with an average call being about four minutes. Using an identification number and a one-minute call-back to the 900 line, the lost pet's status is available for a two-week period.

Dr. Jiff Weber, veterinarian and national spokesperson for the AHA, called the hotline the most comprehensive pet-tracing system ever. Weber also noted that a lost dog or cat is immediately linked with its owner and lost or found pets can be registered day or night. According to the AHA, of the million pets lost each year in the United States, only about 16 percent of dogs and two percent of cats return to their owners.

A portion of the proceeds from the hotline will benefit the AHA, and the balance of the revenues will be used to offset the 800 cost.

The hotline is being marketed through posters distributed to veterinarians and animal care and control professionals. There are also plans to print the 800 and 900 numbers on milk cartons, grocery bags, pet food packages and refrigerator magnets. *Sept. 1992 (active)*

Pet Hotline Still Barking

Last month, the Lost/Found Pet Hotline at 900-535-1515 began its third year of operation. Pet owners call the 900 number to report a lost pet; a separate 800 number takes calls from those who have found a lost pet. The cost of the 900 line is $1.95 per minute.

The hotline is well received by the public, and the 900 proceeds cover the costs of both lines, said Phil Cohen, president of Phil Cohen and Associates, the line's IP. The Tampa, Florida-based company, a producer of educational videos uses its video program for veterinarians' offices to help promote the line. Ads within the video give the hotline number.

Cohen said that when the line began in 1992, he sent a mailing to animal shelters, police agencies and veterinarians to inform them of the line. Research by his company showed that when a pet is lost, owners most often contact their local animal shelter, police or veterinarian. Existing animal welfare agencies support the hotline because it performs a public service, Cohen said.

Typical call counts for the 900 line are 600 a month with the average call costing $4. Sprint Telemedia is the service bureau.

Through virtually identical menus, callers to the hotline's 800 and 900 lines provide their telephone numbers and location by zip code and a description of the animal they have lost. When matches are found, the system dials the telephone number of the caller who reported finding the pet. *Sept. 1994 (active)*

NY Corporation Data Via 900

Increasingly, government agencies are using 900 to raise revenues. The New York State Department of State's Division of

Corporations, is instituting 900-TEL-CORP (900-835-2677) to provide information on corporations doing business in N.Y.

The line began operation last fall using live operators. For $4 per call, callers can verify information on up to five corporations, including correct corporate name, status, address, and names and addresses of any registered agents.

According to William F. Brown, Director of Public Affairs and Information Services with the Department of State, they have a contract with MCI to operate the line. Brown said the estimates are for 150,000 to 200,000 calls per year with a minimum of $500,000 in annual income.

The line utilizes the Division of Corporations' database listing more than two million corporations, mostly in the United States, with some over seas companies listed. *Jan. 1993 (active)*

Private 900 Number Just for Lawyers

Fresno County Superior Court in Fresno, California has instituted a 900 number which attorneys can dial to find out any recent ruling on a certain case.

The number is not available to the general public. Attorneys obtain it from the Fresno County Bar Association and pay $1.98 per minute. The system began designed for only one judge's courtroom, which hears civil cases, but Gary Krcmarik, assistant executive director of Superior Court, said he hopes the system can be expanded to other judges' cases.

With the 900 number, attorneys don't have to travel to court to find out rulings. In addition to freeing up court time, the system is expected to save litigants hundreds of thousands of dollars in legal fees. *June 1994*

Property Assessment on 900

The Nashua, New Hampshire Assessors Office now uses a 900 line for information on property assessments. The line is designed for use by appraisers, developers, bankers, real estate agents, public utilities, attorneys and credit card agencies.

Calls to 900-448-8003 cost $3 for the first minute and $1 for each additional minute. The line is operational from 8:00 a.m. to 5:00

p.m. Monday through Friday. Nashua-based Innovative Telecom Corp. is the service bureau.

Property owners will still receive assessment information free of charge through a local phone number. Mayor Rob Wagner said he has supported the 900 line in order to reduce costs and increase local revenues *Dec. 1993 (active)*

Sex Offender Line Pending

Last month California's Senate unanimously approved legislation to create a 900 application to identify California residents who have been convicted for sex offenses *(AN* July '94). The bill, originally cleared in the Assembly, is now awaiting Assembly re-approval. Residents would call the 900 line to inquire whether or not known sex offenders were living in their neighborhood. Only limited information about crimes would be released, and callers would not be able to find out the address of the sex offender. The legislation also requires the Attorney General to publish and make available annually a list of people deemed to be habitual sex offenders.

Assemblywoman Barbara Alby feels that parents would be most likely to call the 900 line. Proposed call costs are approximately $4 for the first minute and $2 for each additional minute. Any money raised would fund an associated public information program. *Oct. 1994*

900 Number Tracks Molesters

The Child Protection Act, legislation to create a 900 application allowing California residents to access a database of felony sex offenders *(AN* Sept. '94), has been signed by Governor Pete Wilson. Now California residents will be able to find out whether or not an individual is a known sex offender.

Attorney General Dan Lungren, sponsor of the legislation, said, "This (the 900 line) will be the single most important asset for parents to protect their children and prevent one of the most destructive crimes imaginable."

Assemblywoman Barbara Alby, who wrote the legislation, said that the 900 number is expected to carry 6,000 calls per month. Initial profits from the line will return to the state. If the line is profitable, the call cost of $4 for the first minute and $2 for each additional minute will be reduced. *Dec. 1994*

Scherers Wins BBBs

When businesses considered it necessary to block the majority of their phones from accessing 900 lines, they impacted the number of business applications developed and the number of calls to those applications. The Better Business Bureau® of Metropolitan New York, Inc.'s (BBB) 900 line has suffered because of business blocking. The biggest obstacle encountered by this seasoned line is customers who have 900 blockage, such as businesses, said Carrie Getty, vice president. She added that many phone inquiries from businesses were lost after the 900 number started in July 1991.

The mix of inquiries before 900 was 50% residential and 50% business. According to Getty, it is now 70% residential and 30% business.

Sprint originally carried the BBB's 900 number, but they intend to change soon to AT&T and have another number under a broadening of the BBB's system to include other bureaus. According to Getty, AT&T, MCI and Sprint all bid on the BBB's broader 900 system, and AT&T won. Scherers Communications Inc. is the service bureau. Callers to 900-CALL-BBB pay 95 cents per minute, and the average call cost is $3.80.

The New York bureau has long used 900 *(AN*, June '92). Early reports labeled the line successful. This first 900 effort by the BBB must be, because the BBB's national office began adding other bureaus last October when the Philadelphia bureau went online at 900-CALL-BBB, followed by Miami in January and Buffalo in March. The Los Angeles bureau is next, followed by the N.Y. bureau's switch to the new number. The goal is to have as many bureaus online as possible in order to provide callers with a more national database.

In the second week of operation for the Buffalo bureau's 900 line, Judith Peter, vice president, told *AN* the public seemed to accept the line very well. Similar to the N.Y. bureau, the Buffalo bureau provides callers to its local number with information on the 900 number and credit card option.

The N.Y. bureau introduced credit card billing through its local telephone number in December 1991 with a flat fee of $3.85. Getty noted that the credit card option did decrease the 900 volume a bit; but it brought in callers that have 900 blocking. According to Getty, the credit card option brings in approximately 35% of the call volume, but generates 50% of the bureau's pay-per-call income because of the difference in net yield for credit cards and 900. Getty noted that the bureau tends to net 90% of total credit card call revenue; whereas 900 calls provide a net of only 50%.

The N.Y. bureau receives approximately 14,000 requests for information per month from people checking on businesses or charities. That includes written requests. Call-ins break down into approximately 9,000 calls to the 900 line and 3,500 credit card calls. Research with focus groups has given the bureau some insight into how customers regard their 900 line. The findings conclude that 900 users are in favor of the speed that the service offers, and non-users are intimidated by the per-minute charge and feel a reluctance to pay for information if they can still obtain it gratis through a written request. *April 1994*

APWU Uses 900

The American Postal Workers Union will keep its membership informed during labor negotiations this fall with the help of a 900 number. The line, at 900-288 APWU, is designed to give negotiation updates for 50 cents per minute to the nations's approximately one million postal workers.

When the negotiations begin in early September the line's updates will be supplied via an 800 number by the union president. The system will limit a call's total time to under four minutes, making a maximum charge $2 and reducing the union's overall transport costs.

The 900 number is running on AT&T lines. Telecompute Corp. is the service bureau. *July 1994*

State Dept. Uses 900-555
Tax Status Data on Del. Corporations

The Delaware Department of State has launched a 900 number to provide information about the tax status of corporations within that state. Using 900-555-CORP, it will be the first state agency to employ the new AT&T 555 exchange reserved for business-to-business pay-per-call programs (*AN*, Oct. '92).

For a flat fee of $10, the line offers information on the tax status of any of the 206,000 companies incorporated in Delaware. The Department of State receives about 1,000 inquiries per day, said Delaware Secretary of State Michael Ratchford.

AT&T recently dedicated the 900-555 exchange to business-to-business pay-per-call programs carried on its MultiQuest® 900 Service. It chose 555, partly because that exchange is already equated with information. *Dec. 1992*

Traffic Court Uses 900

San Diego's Municipal Court is using a 900 number to provide information on individual traffic cases. The line operates within the county at 900-505-9000 between 8 a.m. and 5 p.m. on the days court is in session. Each call is 50 cents per minute.. *Dec. 1992*

Traffic Violations Info Via 900

San Diego Municipal Court, in October of 1992, began a pay-per-call line on Pacific Bell lines, answering questions regarding traffic violations. The cost of each call is 50 cents a minute. The average call lasts 5-6 minutes, and they receive about 2600 calls a month.

The public's reaction to the service has been favorable, said Kent Peterson, court administrator. At first they assumed attorneys would be the callers, but it is the general public who use it the most. The Municipal Court initially needed special legislation to begin the service.

Currently, they have two lines and intend to use the funds generated by the lines to bring in improved computer support and additional lines. Most problems, Peterson said, "are resolved on the spot."

Things are going so well they intend to expand the service to the civil and criminal areas. In view of the Municipal Court's experience, the San Diego County Board of Supervisors issued a directive last month, encouraging other county departments to provide service via 900, subject to departmental administrative discretion. *July 1993*

In-Depth Profile:
Winning Numbers Hotline
Hoosier Lottery

Serious lottery players who don't want to wait for the scheduled TV broadcast or who don't have ready access to a TV or newspaper can turn to a 900 number to get lottery results when it's convenient for them, instead of being at the mercy of the media's schedule. The Hoosier (Indiana) Lottery has been offering lottery results since the lottery began in October 1989, by calling 1-900-420-CASH for a charge of $0.50 per minute. It is first and foremost a winning numbers hotline, giving callers the results for all of the Hoosier Lottery's games. Scherers Communications, Worthington, Ohio, is the service bureau.

According to Kurt Wise, the public relations director for the Hoosier Lottery, the purpose of the line is to provide better service to the lottery-playing public. In addition to providing winning numbers, the program offers information on how the various lottery games are played and details on where the lottery proceeds go. Nonetheless, Wise indicates that "the vast majority of callers are only interested in the winning numbers."

The highest concentration of calls is every Sunday morning, following the two biggest drawings of the week on Saturday night, for the Lotto Game and Powerball, a large multi-state game. Apparently, quite a few people either don't stay up for the 11 o'clock local TV news or they are out on Saturday night, so the 900 number is a convenient means for getting the winning numbers after that first cup of coffee on Sunday morning. And a regular player who doesn't happen to get the Sunday newspaper can get the lottery winning numbers with a simple phone call. "A good 90 percent of our callers are repeat callers who know the number (by memory)," says Wise.

At only $0.50 per minute, which barely exceeds the cost of running the program, this program was not designed to be a big money maker in-and-of itself. In fact, Kurt Wise was quite explicit in emphasizing that the purpose of this program is to better serve the lottery players. The lottery itself generates some 500 million in annual sales, and the revenue from the 900 program, around $4,000 a month, is nearly insignificant when compared to that generated by the actual lottery sales. "The money from the program just isn't important to us," says Wise.

When asked whether they had ever contemplated actually conducting the lottery via 900, using it as a vehicle for picking a number and exchanging payment (like the experiment by the Massachusetts State Lottery, a program that is no longer active), Wise was quite adamant that they never considered that option, nor do they ever plan to, for a number of reasons. They did not want to offer a system whereby people could play from home, primarily because it would be too easy for children to access the program with no parental control.

A major secondary concern is the impact at-home play via a 900 number would have on the 4,100 retail establishments across the state that are effectively in partnership with the Hoosier Lottery. A 900 lottery line would essentially cut these retailers out of the loop, denying them a substantial amount of revenue and possibly jeopardizing their continuing ability/desire to offer lottery services of any kind. This in turn would have a negative impact on those who do not have access to an unblocked 900 line, possibly limiting their ability to find a convenient place to play the numbers.

The program is actually marketed very little because it doesn't need to be. "The vast majority of the callers are repeat customers who are active lottery players who call over and over again," says Wise. "We get few new callers but very few of our existing callers drop out of the system." Some advertising is done on TV, on the *Hoosier Millionaire* show, a weekly program airing on Saturday nights, but the number is mentioned only once or twice a year. The number is also mentioned on a recording callers hear when calling the lottery office after business hours. As far as any other marketing is concerned, "there really isn't a need for it because the players know the number," says Wise.

The Hoosier Lottery Winning Numbers Hotline has been an unqualified success since its inception when the lottery began in 1989. The fact that it started out , from the beginning, as a pay-per-call 900 line meant that there was little resistance by the public to paying for the information. Had they started out with a toll free 800 line and then switched over to a 900 line, Wise feels that they "would face a large public outcry and a public relations problem. Because we started with a 900 line, it has become part of the landscape."

In fact, Wise would counsel other state lotteries that are considering a change from 800 to 900 not to do so. On the other hand, states just launching a lottery should use 900 from the beginning and there should be little resistance. "The negatives will get lost in all the hoopla, and once it becomes part of the landscape you don't have to worry about it."

Chapter 4

Chapter 5
Professional Services & Advice

This category is not only one of the more obvious applications for 900-number services, it is one of the fastest growing areas, with professionals of all types using pay-per-call as an alternative method for delivering advice and for collecting payment from their clients. The efficiency of the 900-number delivery/payment mechanism now makes it feasible to offer shorter consultations -- even 5 or 10 minutes -- profitably because all the cumbersome overhead associated with accounting for time and accounts receivable is eliminated.

Small companies and even individual professionals can now offer professional advice by the minute over the telephone. For example, AT&T offers a service called Express 900, which is designed specifically for live professional and technical services. For $75 per month (plus $1000 one-time start-up fee, which is waived from time to time), incoming calls to your 900 number simply piggyback onto your existing telephone line (or any line you designate).

Lower cost options are also available. Resellers will pay all the costs for a dedicated 900 number through a service bureau, and then assign extensions to individual consultants and professionals. The start-up and monthly fees can be quite modest using these services. Intermedia Resources in California and Mega900 Communications in Delaware are two such resellers (see appendix E).

Some people use such a 900 number not so much to make money but to screen callers or to prevent people from trying to take up a lot of their time without paying for it, saying words to the effect,"I'd like to discuss your situation in depth, but as you know I earn my living by giving out advice. Would you mind calling me back on my 900 number?" This will cut off those people who pump you for free information, or those who can't seem to get to the point and tend to ramble on indefinitely.

Virtually any kind of specialized information can be efficiently and conveniently delivered using a 900 number, as you will see from the wide diversity of the following programs.

Dial 900 For Tax Forms

Time is getting short. Still, it is not too late to file your tax return. But what if you don't have the right form or schedule? Now, 1-900-860-1040, can save your day.

IdealDial has introduced a new service for the beleaguered tax preparer. Callers to the 1-900-860-1040 number can request two forms per call at an average charge of $1.50 per form.

Paul Zwillenberg, an IdealDial Account Executive, (303) 233-0300, has been disappointed with the results to date. Originally, IdealDial expected the service to generate some thousand calls per day during the tax season. However, start up problems hampered line operations. Zwillenberg looks for bigger and better results next year. *April 1990*

Diet Program For Rural Areas
1-900 USA-DIET Makes A Difference

Phones are user-friendly, anonymous, non-judgmental, and you do not have to live in Los Angeles, New York, Boston, Philadelphia or Chicago to have and use one. They are in towns like Stillwater, Inverness and Holyhokes. Callers in these towns, who have never used, or been exposed to, 900 numbers are using them. They are aware of numbers like 1-900-USA-Diet because of television and print.

Of course, the majority of callers to the number are from areas that represent the major population centers. The program has had the benefit of a vigorous marketing company, Nan Miller/Geer Associates, Inc. from La Habra, Calif. which was able to generate a good deal of free print and TV coverage.

According to Abram Zinberg of USA-DIET, a profile of dieters in small towns who are morbidly obese, (more than 50 to 100 pounds to lose), emerged. Without having to leave their homes these dieters, now have an opportunity to avail themselves of a program with a variety of aspects, including counseling.

Unable to leave their homes because of embarrassment and lacking access to other programs, such as Weight Watchers, Diet Centers, or Jenny Craig; this doctor-approved program offers them a healthful low fat, low sodium, diet, using a computerized interactive program.

Upcoming USA-Diet innovations include: a liquid meal replacement to get participants started and in the habit of calling regularly; a doctor-endorsed program for pregnant women, with the possibility of obstetricians/gynecologists recommending the 1-900 USA Diet number to their patients; and potential joint ventures with diet food companies that have products available in supermarkets. Sounds like Multi-Media Productions Inc., (USA-Diet's parent company) has covered all bases.

The USA-Diet program evolved from a family of multiple professional talents. Sheldon S. Zinberg, M.D. was concerned about the problem of overweight patients. He combined his expertise with his psychologist daughter's knowledge of overeaters' behavior, his son Abram's and his cousin Neil Kligman's interest in computer and telephone technology to produce this multifaceted program. *April 1990*

Live Doctors On Phone

Doctors by Phone, a live 900 application that charges $3 a minute, is modeled after the medical advice call-in shows so popular on radio.

AN believes this may be a winner because, like other 900 numbers, the service takes advantage of the peculiar combination of anonymity and intimacy possible over the telephone.

The service gives people increased and easier access to medical information, said Dr. Thomas Kovachevich, an osteopath who operates the service with a staff of 75 doctors in Manhattan. Questions cover snoring, advice on what type of specialist is needed for a swollen ankle, opinions on whether surgery is needed, and requests for a detailed analysis of a patient's medical record.

Doctor by Phone promises to answer any medical questions, using a medical library, a computer medical database and the backing of an eight-person medical advisory board. The service does not offer formal diagnoses or prescribe medicine over the phone, and is not affiliated with a medical group seeking referrals. *July 1991*

PDR Drug Queries Line

Every doctor's office has a PDR (Physicians Desk Reference). Medical Economics Data publishes this well-respected authoritative volume that contains definitive information for doctors prescribing drugs.

Medical Economics Data is recruiting newspapers across the country for joint ventures that will make information available on the 200 most widely used prescription drugs.

The number is 900-988-5151. West Interactive is the service bureau. Call cost to the consumer is $1.50 per minute with the caller pressing the first five letters of the drug's name to get a 3-to-4 minute recorded message that describes the purpose of the medication, typical dosage, "dos" and "don't's" for taking the drug, common side effects and interactions.

This is a "biggy" for the 900 industry, and *AN* predicts a positive media blitz. *Aug. 1991*

Med Information Line Begins
New York Times Carries 331 Access Codes

A comforting sight to those of us worried about 900 image problems was last month's half page ad in the *New York Times* for a newly launched 900 number, the Medical Information Line.

The president of Strategic Systems, Inc. of Waltham Mass., Ray Novick, was unwilling to discuss whether this was a joint venture with the Times. If *AN* had to venture a guess, it would be a yes.

Modeled after a premium billing service in England launched in 1987, this new application recrafted into American English has 311 recorded topics. The service bureau is Vocall Communication Corp. in New York, and Sprint is the long distance carrier. Colorado Springs, Washington was selected for the test marketing site. It appears all involved have done their homework and Novick was armed with a variety of studies.

The program is designed to please and generate repeat callers. Each answer, varying 3-8 minutes in length, takes up about eight typewritten double spaced pages while the average home encyclopedia has two paragraphs on each topic. The American Academy of Family Physicians Foundation has reviewed and approved each of the messages.

To see what medical topics are covered, you can find out how to get a directory by calling 900 535-3600. The first minute is free and each additional minute is $1.35. The four digit access codes for the messages are printed in the advertising. *Oct. 1991 (active)*

Not-for-Profit Cancer Line

The Cancer Network operates an AT&T 900 telephone line -- 900-RESPOND -- that provides prerecorded information on a range of cancer-related topics and gives callers the option to talk one-on-one with a cancer survivor. Revenues from the call ($2.95 for the first minute, and 95 cents for each additional minute), keep the nonprofit group in business. *Dec. 1991*

Arthritis Hotline Operating
Ads in TV Guide Help Promote Line

Over the years we have heard from people who are involved with a subject and have collected information that they are anxious to share with others -- using 900.

Well, Ray Chrisitie has done just that. A jazz musician, his career was cut short by severe arthritis. Ray spent two years researching the medical and scientific data and organizing it into an interactive audiotex service.

Chrisitie reported, "I am reasonably pleased with the numbers of calls we have been getting. The advertisements in *TV Guide* have been working real well, considerably better than cable TV spots." Chrisitie is looking for media or marketing partners.

The Arthritis and Back Pain Information Service uses software developed by Philadelphia's Gateway Telecommunications. Priced at $1.75 for the first minute and 95 cents each additional minute, 900-88-Help-Now is a menu-driven phone program that enables callers to research specific areas pertaining to their individual needs and concerns.

"Although a cure for arthritis has proven elusive, vast amounts of treatment information are available," said Chrisitie. He feels the line's low price is a partial explanation of why it's doing well and explained, "Arthritics read the TV Guide because it helps get through long days and nights." *Dec. 1991*

Healthy Baby Line

The March of Dimes Birth Defects Foundation has chosen the 900 technology as one of the ways to help the nearly one-third of all pregnant women in America who do not receive early or adequate prenatal care.

The Healthy Baby Message Line, 900-288-Baby, gives prerecorded messages on healthy lifestyle before and during pregnancy; early pregnancy detections, and the three trimesters of pregnancy. The charge to callers is 75 cents per minute. *May 1992*

Healthy Baby Line Growing

The March Of Dimes Birth Defects Foundation expects steady growth for its Healthy Baby Message Line *(AN*, May 1992), especially after the upcoming national promotion. Women calling 900-288-BABY can hear prenatal advice and obtain the phone number of the nearest March of Dimes chapter, from which they can request prenatal educational material.

The 75-cents-per-minute line is also accessible to rotary dial users, an important feature. "We wanted to make sure that we could reach all of the country, not just those people with touchtone phones," said Holly Holmes, March of Dimes Birth Defects Foundation's special assistant to the director of national promotions. Holmes told

AN that studies show 38 percent of consumers in the United States have rotary dial telephone service.

The line's sponsor is Warner-Lambert Company's e.p.t., an in-home early pregnancy test. Through the national promotion, the Healthy Baby Message line number will be listed on both e.p.t's point-of-purchase brochures and the instructions inside each kit. Other promotion is being carried out at a grassroots level, Holmes said. The March of Dimes 130 chapter offices and 250 division offices are placing press releases and ads in local media.

According to Holmes, Warner-Lambert gave a donation to fund the startup of the line, and the cost to callers covers the line's expenses. "We stress that this is not a fundraising venture," Holmes said. "We don't make any money on this service."

The average call has been under three minutes at $.75 per minute. According to Holmes, the line's call counts have remained in the double digits per week, with calls coming from all areas of the country. Advanced Telecom Services is the service bureau.

As part of the March of Dimes Campaign for Healthier Babies, the line is intended to both educate and encourage pregnant women to get into prenatal care as soon as possible. *Oct. 1992*

Claims Plus Filed Via 900

Doctors, dentists, pharmacists, clinics and other health care professions will now be able to electronically file insurance claims -- regardless of the insurance carrier. By dialing a single 900 number and using an interactive voice response system, proponents say they will save money and speed payment to the care provider.

Health care providers register to use the Claims Plus, Inc. system by placing a $25 call to 900-370-8887. After registering, a doctor's office files the individual claims by dialing 900-773-7272. For $2 for the first minute and $1 for each additional minute, the caller inputs the claim data.

Upon completion of each call, Claims Plus formats the claims data and electronically transmits the claim to the appropriate insurance company within 24 hours of receipt. An average claim takes about three minutes to input. Multiple claims can be processed on the same call. *March 1992*

High Risk Insurance Line

The *Insurance Journal* has established a 900 number to help insurance agents locate where to place high-risk items. The cost to access the system is $2 for the first minute, and $1 for each additional minute, although blocks of time can be bought at volume discounts of up to 25 percent.

Callers to 900-97-Markets get a voice prompt to guide them through the menu or can receive a User's Guide to speed the process. *May 1992*

Sprint Offers Legal Advice

Pre-Paid Legal Services Inc. of Ada, Oklahoma and Sprint Gateways of Overland Park, Kansas, a unit of US Sprint, will be making legal advice available for $9 for the first minute and 50 cents for each additional minute through a new service called Justice 900.

Pre-Paid Legal Services has 18 years experience providing legal services on a subscription basis, and their attorneys work through independent law firms under contract with the company to provide specific types of legal services.

Callers to 900-230-Justice are given a toll-free number and an access code for an attorney in their home state. *July 1990*

Law Form Fax Service

Automated Fax and Voice Solutions, Inc. (AF&VS), a N.Y. -based service bureau, in conjunction with Julius Blumberg, Inc., announced that 750 N.Y. and national Blumberg law forms can be received in minutes by fax. AF&VS claims this is the first commercial application of a digitized graphic database to transmit law forms instantly.

Customers receive a form by calling 900-Fax-A-Law and entering their fax number and catalog form number. The charge for the call is $12.50 and faxes of catalog numbers can be received by calling 800-666-LIST.

Dean Garfinkel, president of AF&VS, predicts this application is a winner. He explained, "This company is the largest private publisher of law forms. Just about every transaction executed in N.Y., from tenant leases to divorce filings, requires a Blumberg law form. Our application allows instant easy access to these forms. These are the first of many forms we intend to make available by fax. Law form libraries for other states are next." *Feb. 1992 (active)*

New Legal Service Starts

A new Call A Lawyer (900-820-3005) service has started in Houston. The information is live from the attorneys with the average call taking 8-12 minutes at a cost of $3 a minute. Although not necessarily for the complicated case, Attorneys Steve Brownhill and Doug Wheat believe it is helpful for the average citizen with legal questions. *Dec. 1990 - Jan. 1991*

Live 900 Works Well
Industry A Mecca For The Modest

Jeffrey Powers, who believes he is a marketing whiz, has a live line that he answers from a portable phone that goes wherever he goes, including the bathroom. Previously, 900 lines billed to assist people getting into the audiotex industry have been recorded messages.

Powers said: "The callers are impressed with the efficiency of the system. It's instant advice anytime they want. Plus you don't need to do the billing. It's all done through the phone company."

He advertised the line through the *L.A. Times* classified until they caught on it was a 900, and then he had to switch to display, which was not as effective. He advertises through the new business opportunities markets as well. His advertising budget was $1000 a month, is now down to $500 and will shortly be zero because of call backs and referrals.

Marketing isn't the attraction. Powers believes it is his independent objective opinion which is more appealing than a recorded message.

When he began he called AT&T, Sprint and MCI. They all told him it was impossible. "They said 900 lines were for recorded message not talking to people." Despite the turndown, today Powers sees MCI as the leader in the field.

Powers worked with Telesphere and chose 900-9-Mentor as his number with a call-cost of $3 a minute. Operational since March, he made $700 the first day and now gets 240 calls a month averaging 24 minutes each.

Powers' Westlake Village, Calif. company, is Jefco/Checkpoint. *Oct. 1990*

Live Lines
Phone Therapy Offers Refunds

TalkPlus, a new application, offers short-term therapy. Calls are answered live by trained therapists. Phone therapy is a logical extension of the trend in psychology toward brief therapy, says Dr. Allen Fay, a New York psychiatrist who started the program.

If callers are dissatisfied in any way with the service, a refund will be issued upon receipt of a copy of the customer's phone bill. The cost of a call to 900-346-8855 is $2.80 per minute. *Oct. 1991*

900-Druggist

Consumers will call 900 numbers to get information on health topics, especially when the information relates to prescription drugs *(AN,* February 1993). 1-900-7-Druggist is a new 900 line offering callers personal consultations about their prescription drugs.
Debbie Avant, an IP and registered pharmacist, began the line in May. She handles the calls from 9 a.m. to 6 p.m. Monday through Friday. Calls costs $2.99 per minute. Denver-based IdealDial is the service bureau.

Avant sees 900 as a way to reach the widest segment of the public with prescription drug information. Having worked in the retail pharmacy field until injuring both arms several years ago, Avant knows that consumers often have questions about prescriptions that pharmacists don't have time to answer in the hectic retail environment.

Working with SFP, an advertising/public relations agency, Avant is marketing her line through newspaper articles, direct mail and print ads, such as one running in Prevention magazine, a health publication. She is also considering marketing the line as a joint venture to large companies who could offer the service through 800 or 900.

Avant can answer callers' questions on a drug's side effects, benefits, dosages and interactions with other drugs. She shows confidence about the importance of her line's information. "Medical statistics show that accurate drug information saves thousands of hospitalizations and doctor's office visits per year." *July 1993*

Masters & Johnson Go 900

Inside Edition, a TV show seen in over 120 markets in the US and Canada, introduced the Masters and Johnson National Human Sexuality Information Line -- 900-463-4636.

Robert Goot, president of New Wave Communications, Inc., said the number of calls they had received was proprietary, but that he was pleased, saying, "The last two weeks have been excellent."

New Wave is currently planning the marketing of the line, anticipating their start-up in March, when *Inside Edition's* restraint is no longer in effect.

The Masters and Johnson Institute is one of the world's foremost sex research clinics. The Information Line enables callers to hear factual information in the privacy of their own homes. *Jan. 1992*

Affirmation Line

The information dispersed through 900 meets many needs for callers, and now one line gives callers a verbal pat on the back. The 900 Affirmation Line at 900-420-3709, Ext. 102 gives callers positive statements about what wonderful people they are. Calls to the frequently updated line cost $2 per minute.

The line's IP, Bart Walsh, a Portland, Oregon licensed clinical social worker, got the idea for the line from his work. He realized what people long to hear most. *Nov. 1993 (active)*

Non-Judgmental Sex Line
Info Pays Off For Center-- Phones Ring

Sex and 900 are a good marriage -- as proved by the Institute for Advanced Study of Human Sexuality in San Francisco. The need for accurate, non-judgmental sex information led Marguerite Rubenstein and her colleagues at the institute to set up the telephone line.

According to Rubenstein, they have received 10,000 calls in the first three months of service. The number, 900-226-4327, costs $2 for the first, and $1 for each additional minute. Answers last two to four minutes and cover topics under the general headings of How to Have Safe Sex, Sexual Options, How to Function, What Can Go Wrong, Sexual Abuse and Am I Normal? *Nov. 1990*

Sex Therapy Via 900

Dr. John Limbert, a sex therapist, believes a 900 service he has started is invaluable for people too traumatized to seek out information any other way.

"I see it both as a health venture and money-making operation," he said. "It's very much in line with my philosophy working as a sex therapist -- recognizing that many of my patients are uncomfortable about approaching someone to address their sexual problems."

For $2 a minute, callers to 900-Choices reach a bank of recorded messages about impotency, sex health, sexual addiction and other related topics. *May 1991*

"Family's" Line Advises Parents

Readers can add to their list of 900 applications that project a positive image for the industry. *Good Family Magazine's* new 900 number offers parents advice on more than 20 subjects dealing with child care.

Callers to 900-884-KIDS pay $1.99 for the first minute and 99 cents for each additional minute. Kathy Boice, director of new marketing development, said that the 900 line follows the company's philosophy of providing information for the gratification of families and encouragement of the family unit. Advanced Telecom Services is the service bureau. *July 1993*

Attorney's Line Summons
Questions and Angry Callers

George Anagnos, a New Jersey attorney, is an IP, which is not unusual. What is unusual is the two diverse uses he has for his 900 line. Callers to 900-773-8245 can ask for legal advice or vent their hatred of lawyers for $5 a minute. For the line's latter use, Anagnos runs advertising saying "Hate lawyers? Curse out a live one." Anagnos says that he normally empathizes with callers who express frustration over lawyers.

He began his 900 number for legal advice last January, aimed at callers from New York and New Jersey. The "hate lawyers" side of the line came later when he was searching for another use for the 900-service after failing to get many calls for legal advice. *Sept. 1992*

Cop Talk-Police Procedures
Law Enforcement Questions Answered Via 900

The diversity of information available through 900 continues to widen. Cop Talk, which debuted at 900-835-0911 last fall, gives callers a chance to ask an off-duty or former police officer questions about law enforcement.

According to Mark Miller, manager of administration and publications of Cop Talk's sponsor, Security Strategy Systems, Inc., the line gives the public information on police procedures they may not be able to get from their local police precinct because of a lack of staff or time. Miller stressed that the service does not offer legal counseling, nor will Cop Talk agents help callers whose questions indicate interest in committing a crime.

Half of the callers ask questions relating to traffic tickets, Miller said. Other calls cover a wide range of subjects from handguns to parents' concerns about a child's behavior. Callers pay $2.49 per minute. Those using pay phones can dial 800-753-0911 and bill charges to a credit card.

Caller response has been good. Miller said that the majority of callers "are thankful we're here and feel good about the information they're getting."

Radio has been a mainstay of the marketing effort for Cop Talk. When *AN* spoke to Cop Talk, Glenn Woodworth, the line's founder and president and CEO of Security Strategy Systems, was in the midst of a heavy schedule of radio interviews. Radio ads have also been used.

The Troy, Ohio based Security Strategy Systems, Inc. has 40 off duty or former police officers from around the Troy area acting as

Cop Talk agents. Reception from the local law enforcement community has been good, Miller said, with the Cop Talk agents enthusiastic. Also, a number of Fraternal Order of Police chapters have expressed support for Cop Talk.

A book "The Bare Facts About Bear Traps," is a spin off of the line. Due out this month, the book contains advice relating to traffic tickets. *Feb. 1993*

Counseling Line

Every operator of a live 900 line knows that it can take a few months to work the glitches out of the line. Psychotherapist Dr. Frank Husted of the Phoenixville, Pennsylvania-based TELE-HELP is no exception. His counseling line at 900-225-HELP experienced some organizational and technical problems after its beginning in October. Husted, however, is taking the early problems in stride and is looking forward to the line's growth.

Husted sees telephone counseling services as the wave of the future for mental health care because of the privacy and anonymity the telephone offers. He told *AN* that frequently in his work as a psychotherapist clients want to talk to him over the phone before making an appointment, something which helped steer him toward 900. However, he stressed that TELE-HELP does not offer therapy over the phone, but counseling and referral for face-to-face therapy if needed.

Last month, the $3.99-per-minute line had 20 counselors working as independent contractors. Husted said that he would like to limit the number of counselors to 300 to 400 throughout the country. Counselors include social workers, psychologists, and substance abuse and victim assistance counselors. Husted noted that an ad for counselors he ran in the *Los Angeles Times* drew 200 responses. "People are interested in working in this medium, and I'm excited that it's going to be going places," he said.

Advertising for the line has included print advertising in *Cosmopolitan*, a group of local newspapers in New York and a singles magazine. A 30-second television spot debuts this month on the national cable channel *Prevue*.

Pennsylvania-based Information Service Bureau Corp. is the line's service bureau. *Jan. 1994*

Drug Info Line Two Years and Still Going Strong

Marketing is undoubtedly the lifeblood of the 900 industry, but Ask the Pharmacist, Inc. does not have to invest time or money in marketing its 900 line. As it goes into its third year of operation, Ask the Pharmacist, at 900-4200-ASK, continues to gather enough interest from the medical community to keep it well publicized.

"We're still a real point of interest," said Mary Lynn Bell, president of Ask the Pharmacist, Inc. Among the recent publicity is an article in the Harvard Health Letter. The live line, which began in Feb. 1991, answers consumer questions on medication. Callers pay $1.95 per minute to consult personally with a registered pharmacist. Another factor that relieves Ask the Pharmacist of actual marketing efforts is that, since late 1991, pharmacy-related companies have used its lines for 900 or 800. "We supply the information and expertise," Bell said, "the client does the marketing."

Ask the Pharmacist uses AT&T lines and its own phone equipment. Bell said that she regrets not having spent more money on state-of-the-art phone equipment when the line first started thus avoiding the purchase of a second set of equipment.

According to Bell, the line is doing well. "It's growing," she said. "So many more people are aware of the need for drug information." Callers ask about a medication's side effects, interaction with other drugs, and proper dosage. "But, we do not diagnose, ever," Bell said.

The line is staffed by up to 60 pharmacists and has access to the most up-to-date information on medication from several software packages, as well as books and periodicals. *Feb. 1993 (active -- see in-depth profile at end of chapter)*

Florida Attorneys Begin Law Line

Count in another legal line. The Law Line, at 900-884-LAWS, a live line run by a group of South Florida attorneys, allows callers to consult licensed attorneys.

The cost is $4.99 per minute, with the first minute free. The line's staff answer questions on numerous legal issues, such as criminal cases, real estate and traffic violations. Callers are referred to other sources if staff cannot help. *June 1993*

An IP's Dream Come True!
Counseling Line Gets Print & TV Publicity

Every 900 IP craves publicity for his line, and a new telephone counseling line achieved enough initial publicity when it went online last December to become an IP's "dream come true."

On the same day the Summit Solutions Line went up on 900-680-6338, *USA Today* ran a story on it. The next day CNN covered the line, which gives callers the chance to talk to licensed therapists. "We've been very fortunate," said Dr. Robert Marrone, director of professional services with the Summit Services Network. Marrone added that initial marketing efforts included press releases and some infomercials that began airing this January in six cities-- New York, Raleigh, N.C., Chicago, Seattle, Los Angeles, and Philadelphia. The advertising and public relations firm Torme & Co. coordinated these efforts.

According to Marrone, calls doubled after the CNN story and went up still higher after the infomercials began running. When *AN* spoke to Marrone, in the third week of the line, he had call counts numbering in the thousands.

The $2.99-per-minute-line offers callers help with life's issues and difficulties. Marrone noted that the line's approximately 300 counselors, including clinical psychologists, marriage and family therapists and clinical social workers, are all licensed and trained in offering brief counseling techniques. The average call is 15 to 18 minutes long, he added.

The line kicked off operation with a "post holiday blues special" price of $1.99 per minute. The regular price went into effect Jan. 6.

Summit Services Network also runs the Summit Solutions Line on 800 with the same number, and callers pay for the service by credit card. The immediacy of 900 as well the convenience of 900 billing led the company to use it along with 800, Marrone said. West Interactive is the service bureau.

He noted that the company also publicizes the 800 line in its infomercials and other advertising and publicity. In addition to the infomercials, future marketing will include television and magazine advertisements. *Feb. 1994*

Legal Information Via 900

Attorneys looking for opinions rendered by the Calif. Appellate Court and the Ninth U.S. Circuit Court of Appeals can access that information by dialing 900-860-6066.

For $2 a minute, with the first minute free, Attorney's Briefcase, Inc., an Oakland-based legal research software company, has launched this new modem-to-modem service. The legal decisions and statutes are indexed by topic, citation and date, and are generally available within 48 hours of their release.

The system has a "Week in Brief," listing statutory changes and legal trends. There are search and retrieval commands with user-friendly indexing, and callers have the option of downloading text onto a hard disk for later review. IdealDial is the service bureau. *Feb. 1993*

K-Mart Healthcare Line

Late last year K-Mart introduced a telephone co-op line aimed both at enhancing the image of its pharmacies and increasing prescription and over-the-counter drug sales. The K-Mart Pharmacies Healthcare Line operates with an 800 number, 800-77-HEALTH.

Dispensing basic information on approximately 100 health topics, the line offers callers a $5 discount for prescriptions filled at a K-Mart Pharmacy outlet. Fifty participating drug manufacturers provide coupons for the callers. The service bureau is Wayne, Pennsylvania-based Advanced Telecom Services, Inc.

Test marketing began in November in Columbus, Ohio; Indianapolis; and Tampa/St. Petersburg, Florida. A direct mail campaign sent directories listing the line's extension numbers to 1.7 million homes. K-Mart Pharmacies in those areas also distributed directories.

Callers can listen to a minute to minute-and-a-half recordings on healthcare topics. The line's preamble stresses that it is not intended to diagnose health problems or be a substitute for a caller's physician, and that callers should consult their physicians for serious problems.

With this sophisticated application, callers, based on their responses, are mailed a written copy of the health topic message they listened to, coupons for a related product and the K-mart Pharmacy coupon. They also receive an additional directory to give to friends and family.

K-Mart carefully controls the names and addresses collected by this line. Only the brand sponsor of the product the caller expressed interest in receives that name and address. Brand sponsors pay for their participation on a per-call basis. *Feb. 1994 (active)*

Line Begins as Joint Venture

The International Network of Children and Families (INCAF) has begun the National Parenting Hotline, a live parental advice line, inspired by response to a Florida television news show's recent five-part segment for parents. A two-hour call-in period after each night's airing brought a deluge of calls to childrearing experts present at the station.

The news series based on the nationwide course, "Redirecting Children's Behavior" (RCB), ran in February. There were more than 300 calls, with another 500-700 callers unable to get through. This local response showed the INCAF that there was enough interest to warrant reaching parents nationally through 900.

"We feel it's a tremendous service to parents and children, and there's a real need out there," said INCAF member Bob Hoekstra. The hotline, begun last month, is joint ventured with newspapers. Calls are $3.95 per minute.

Participating newspapers provide advertising for the line, and the INCAF is producing an infomercial. They anticipate publicity through radio and parents' magazines, said Barbara Abramson, owner of Barbara Abramson Marketing and Consulting Services, which is marketing the line.

The INCAF was negotiating with more than a dozen newspapers at the time *AN* spoke to Abramson. Delray Beach, Florida-based ICN is the service bureau.

The line will connect callers with any of several hundred qualified RCB instructors on standby. These instructors can offer advice on a wide variety of childrearing problems, from sibling rivalry to school problems. *July 1993*

Mag Tests 900
Architectural Digest Gets Closer Via 900

Architectural Digest, a leading magazine of interior design, used the last three months of 1992 to test a $4, two-minute 900 call. It selected Sprint Telemedia as its service bureau.

This was the magazine's first experience with 900, and it is currently evaluating the results, said Larry Karol, director of administration. The line was aimed at the magazine's readers and marketed primarily through the advertising on its pages, Karol said. "This was a way to bring us closer to our readers."

The magazine's editor-in-chief, Paige Rense, read the messages, which offered practical decorating advice from a different designer each week. *Feb. 1993*

Medical Info Line One Year
and Line Still Going Strong

One year after its launch, the Medical Information Line *(AN* Oct. '91) is going strong, attracting approximately 20,000 calls per month. Callers to 900-535-3600 can access any of 350 recorded messages, each on a different health topic, for $1.75 per minute with the first minute free.

Operated by Strategic Systems Inc. of Waltham. Mass., the line carries easy-to-understand messages reviewed and approved by the American Academy of Family Physicians Foundation. The topics range from alcoholism to ear wax to heart attacks. The service bureau is Vocall Communications.

According to Strategic Systems Inc., the most frequently accessed topics include women's health issues, sexual matters, mental health and cancer. The typical message lasts five minutes, and each contains referrals to related topics.

The marketing efforts for the line have included a half-page ad in the New York Times used when the line was new. Other efforts, as *AN* staff members have seen, include information cards distributed from bulletin boards at grocery stores. *Dec. 1992 (active)*

Parents Leave Messages
Center Offers Two Lines: Same Data Bank

The National Parenting Center is an organization with a firm grip on 900. The center cosponsors 900-535-MOMS with Parenting magazine and solely sponsors 900-246-MOMS.

Both lines use the same service bureau, LO-AD Communications, and offer parents recorded messages by experts in childrearing on such topics as feeding, education, health and more.

The center's own line began in 1989 and the venture with Parenting a year later. The cost of calls to either line is $1.95 for the first minute and 95 cents for each additional minute, with the average call about three minutes.

The menu offers callers a choice of seven age groups -- pre-natal, newborns, infants, toddlers, preschool, preteen and adolescence -- five messages for each group. The messages are about one minute in length and change daily, said David Katzner, president of The National Parenting Center.

According to Katzner, one of the most utilized parts of the system is the messaging feature. Callers can leave a question and receive a personalized reply from one of the nine-member panel. That panel is a major selling point for the 900 lines, Katzner said. "I believe a couple of parenting lines have come and gone. What sets us apart is our panel of experts." Katzner added that they are well-known. "People feel comfortable with our experts."

The 900-535-MOMS line is marketed through advertising in Parenting, and the center markets the other line among its membership. *Feb. 1993*

Mental Health Not Just Another Line

A major health services company has launched the Mental Help Line created to permit callers direct access to licensed therapists. The Norfolk, Va.-based First Hospital Corp., an operator of mental health services across the nation and manager of hospitals and HMOs, has added a new unit to its mental health division to operate the line.

Using West Interactive as its service bureau, First Hospital Corp. is test marketing the $3.99-per-minute line in Chicago and Los Angeles. The company is using only experienced practitioners as counselors, all of whom undergo both a training period and a readiness assessment. The project also has built-in quality control via anonymous callers from an outside company.

The line uses "Solution-Focused Therapy" in which counselors will suggest a specific task to bring a caller closer to solutions to his problems. Counselors will strive to limit calls to 10 to 15 minutes. *Sept. 1993*

Pharmacy Line

As it enters its second year of operation, Call Our Pharmacist, a 900 application offering prescription drug information, is building momentum slowly and reaching an untried audience for pay-per-call services -- car phone users.

The $1.99 per minute line at 900-90-DRUGS offers both recorded messages and live consultation with a pharmacist, plus printed material mailed out at no extra charge. Francis J. Muno Jr., president of Bartlett, Illinois-based Patient Prescription Information, Inc., the line's parent company, told *AN* that as of this month a new arrangement with CellularOne in the Chicago area allows CellularOne customers to access Call Our Pharmacist from their car phones by dialing "*RXX." Callers hear the line's audiotex recordings free of charge and can speak to a pharmacist after giving credit card information for billing.

CellularOne will market the service to its 1.5 million customers. Other marketing for Call Our Pharmacist goes on through networking with poison control centers, hospital emergency rooms and consumer health organizations, Muno said."Our company tries to reach consumers in the places they go for general information, such as at poison control centers, where one-third the calls are information requests.

Callers to the 900 line access the line's recorded messages by punching in the first five letters of the name of a prescription drug.

The two- to three-minute message tells the caller what the drug does, and gives dosage instructions and general cautionary information. The line's information is updated constantly, Muno said. Muno spent more than a year preparing to go online. His company does not use a service bureau and owns all its own equipment.

Patient Prescription Information also operates an 800 number, which utilizes credit card billing to allow callers to speak to a pharmacist. The 800 line also gives the 900 number and describes its functions. *March 1994 (active)*

Sex Info With the Masters Touch

Ordinary people with imagination can "make it" in 900, as the experiences of Dave Russell, a Florida information provider, show. Russell's efforts have been the driving force behind the Masters & Johnson Institute's 900 sexuality hotline launched this month. This $3.99-per-minute live line features well-educated counselors personally trained by Masters' staff.

Russell contacted Dr. William Masters after a friend of his had the idea for the 900 line. He told *AN* that it took him more than a year and a half to put the pieces together, including getting AT&T's approval. After all, this is a "live sex line"! Russell cited West Interactive, the line's service bureau, for helping.

Russell's line has had amazing publicity without the help of a public relations firm or a single press release. According to Russell, Masters was at the American Psychological Associations's conference in Toronto in August and mentioned the program. The press picked up on the story, and the phones started ringing. The line is using print advertising and two 30-minute infomercials, airing since October 1. This is not Masters' first venture into 900. An earlier recorded line *(AN* Jan. 1992) was carried by Sprint Telemedia.

Counselors on duty at 900-454-9922 screen callers to verify their age and keep calls down to 30 minutes maximum. About 500 counselors will handle the line, working out of Masters' St. Louis headquarters. All have advanced degrees.

In an article in the *Los Angeles Times*, renowned sexologist Albert Ellis said that Masters' reputation should help elevate this 900 line above other programs.

The two 30-minute infomercials are in a talk show format featuring a panel of experts. They were produced by Century III at Universal Studios, using facilities at Channel 35. Certified sex therapist Judith Seifer hosts the talk show format. *Oct. 1993*

Telephone Counseling

Prompted by the trend toward short-term or single session therapy, mental health professionals are increasingly offering telephone counseling.

One such line is Tele-Counselor, sponsored by Mich.-based Macomb Family Services at 900-740-CARE. For $2.45 per minute, callers can talk with an experienced therapist. The line operates from 8:30 a.m. to 11:30 p.m., staffed by six counselors. The average call is about 20 minutes.

Program Director Sherry Carson said that the treatment form through the line is brief counseling. Carson stressed that the line is not trying to replace traditional therapy or psychiatry. Counselors with professional specializations deal with terminally ill people and victims of sexual assault.

Tom McWhirter, founder of another line, Psych-Line, said that telephone counseling works. According to McWhirter, some suggestions, insight and support can really help people. *June 1993*

Save Your Pet's Life

Has your pet just ingested some poisonous substance? Get quick help from The National Animal Poison Control Center by calling 1-900-680-0000 and talk to NAPCC licensed veterinarians and toxicologists for $2.75 per minute. *(active)*

Healthy Baby Line Ceases

After two years of operation, and a great deal of positive press, the March of Dimes Birth Defects Foundation disconnected its Healthy Baby Message Line. The 900 line operated in partnership with the Warner-Lambert company, and was promoted through the company's e.p.t in-home early pregnancy test kit *(AN,* Oct. '92).

A March of Dimes spokesman told *AN* that when the partnership with Warner-Lambert expired in June, the line was stopped. Point-of-purchase brochures and a coupon with the e.p.t kits listed the 75-cents-per minute 900 line that gave callers prenatal advice. *October 1994*

Safe Sex Education Line

A Pennsylvania man has begun a 900 line designed to give callers' opinions on the effectiveness of condoms and alternatives to preventing HIV infection, such as abstinence and monogamy.

The Safe Sex Education line at 900-4-INFO-VG costs callers $1.95 for the first minute, plus 95 cents for each additional minute. IP Stephen Gabb of Gabb's Educational Enterprises reported that marketing planed for the line included advertising the line on PAT buses. *October 1994*

Speeding Ticket Advice

Callers who need help with handling a speeding ticket can dial 900-448-9948 and hear advice on contesting that ticket. The line covers four landmark speeding ticket cases and a list of a dozen things an officer must produce in court so a ticket "sticks."

IP Quinn Williams, who is part of the security force at a U.S. Naval Air Station, calls himself a traffic cop with a conscience. *October 1994*

900 Medical Line In Canada
"Things Look Good Up There"

Medical lines have carved a strong niche in the United States 900 industry, and now a medical line is among the first applications up after Canada's 900 launch *(AN*, May '94). MedQuest Ltd., in partnership with the medical publishing unit of McLean Hunter Ltd., launched MedQuest at 900-451-3000.

B.F.D. Productions, through its Calgary bureau, is MedQuest's service bureau. The application uses AGT lines, said Bruce Dyer, B.F.D. president. He noted that his bureau felt that AGT was the most technically advanced. Gaining approval from Stentor on the MedQuest application took three weeks, Dyer said. Dr. Tibor

Harmathy, president of MedQuest, Ltd., noted that MedQuest was only the second Canadian 900 application to be approved by Stentor.

As of last month, B.F.D. had 40 applications up and running in Canada with many more pending. "Call counts, overall, are better than anticipated," Dyer said. "The number of calls is higher per capita than the United States. Things look good up there."

George Harvie, in charge of audiotex services for AGT, said that 70 percent of its initial programming was psychic, datelines and horoscopes.

MedQuest began operation July 11, and logged 4,000 - 5,000 air minutes by the end of that month. Callers listen to messages on any one of 150 medical topics for $1.95 per minute. At least 50 more topics will be added in the near future, said Dr. Harmathy.

In a joint venture, partner McLean Hunter markets the line in its publications, *Chatelaine, McLean's, Parents, Today* and *Modern Woman*. Ads list the line's topics and the codes for each.

A group of 28 family doctors researches and prepares MedQuest's information. The messages, each about three minutes long, present basic information aimed at the average layman. *September 1994*

Dial-A-Detective

Computerized detective work hits the 900 industry. The Detective Hotline at 900-860-0660 allows callers to speak to an "investigative advisor" and find out how to go about locating whatever public information on an individual they need. Advisors tell them what information sources they can use. Calls cost $3.99 per minute, and for an additional fee of $30 to $40, the advisor will do the search, normally by computer. The line is not designed to be a full-scale investigative service, but a link for callers to information computer databases.

The line began last March, tested first in Dallas and Philadelphia before going part-time nationally. Its parent company, MPC Telcom, plans to conduct market research before offering the line 24 hours a day. In its first two months, the Detective Hotline's 40 advisors handled 275 to 300 cases, ranging from background checks on a nanny to a search for an old friend.

The advisors all have police or investigative backgrounds, and include retirees from the Secret Service and FBI. Part of their training for hotline work stresses quickly determining what information the caller actually needs in order to keep call lengths down. *July 1994*

Tele-Lawyer Still Going Strong

This is the oldest and longest running live professional advice program in the country, according to founder Michael Cane, based in Huntington Beach, California. "This was certainly the first live professional consulting service (on 900)," says Cane, "and it may have been one of the first live programs of any kind." Cane is well known in the pay-per-call industry for his professional consulting and for his numerous articles in the audiotext trade publications. He is one of the pioneers in this industry, and has contributed significantly to its growth as a legitimate and respectable information delivery medium.

Tele-Lawyer was established in October of 1989 on 900-446-4LAW at $2 per minute. This is an MCI line that is still active today. Shortly thereafter, Cane launched another line, 900-TELE-TAX, which began in January 1990.

The Tele-Lawyer lines are staffed by licensed attorneys, mostly in the State of California. Although the tax line was initially staffed by experienced tax preparers, it became apparent early that many of the tax questions were legal in nature, and that the tax attorneys on staff were able to handle them. The same staff now handles both programs. Many other telephone numbers (both MCI and AT&T) have since been added to the program, including 900-TELE-LAW. After January 1990 the call charge was allowed to increase to $3 per minute, where it stands today.

In the middle of 1991 Cane launched yet another line, this one a consulting line specifically about the 900-number business, targeted at those who are starting up a pay-per-call program and need advice from an expert who is not selling anything. The number is 900-835-2529 and the charge is $3 per minute. Although Cane handles few of the Tele-Lawyer calls anymore, he still handles most of the 900 consulting calls.

In marketing the Tele-Lawyer program, Cane will occasionally run cable TV ads, concentrated in short "blitzes," to generate awareness and to gain new customers. The program gets free publicity from many legal publications that will mention the program, and they will even get referrals from district attorney's offices. The vast majority of the callers, however, are repeat long-term customers. "We have captured a large portion of the market just by being out there for five years, getting a lot of word-of-mouth exposure," continues Cane.

According to Cane, Tele-Lawyer is the only attorney many of its customers have. Because the program covers every major legal category, with a staff of 36 attorneys, many people really don't need to have another lawyer, and they prefer the economy of being able to pay by the minute for instant advice. The service offers all the specialists found in big, expensive law firms. In many cases a personal relationship will develop between the client and a specific attorney and they will be on a first-name basis. One of Cane's clients, a real estate developer, has even discussed his marital problems on the line!

Tele-Lawyer will typically have three or four attorneys in the office answering the phone on an average day. Some of the specialists, however, such as the immigration and copyright attorneys, work off site from their own offices. The calls are simply forwarded to these attorneys.

Cane's experience with Tele-Lawyer has resulted in his writing a series of books addressing the most frequently asked legal questions in America. Titled *The Five Minute Lawyer* series, the books, published by Dell, are scheduled for Spring 1995 release. Cane went through his Tele-Lawyer database and determined that the hottest topics are debtor-creditor rights & bankruptcy, divorce, estate planning and taxes. He wrote a short book for each topic covering the most frequently asked questions he has fielded over five years on the Tele-Lawyer line.

A measure of how satisfied his callers are is the low rate of uncollectibles experienced on the Tele-Lawyer line. According to Cane, AT&T is always remarking that his program ranks among the best in terms of low uncollectibles, usually measuring under 1%. This program is obviously filling a need by delivering top quality service. *Jan. 1995*

In-Depth Profile:
Ask the Pharmacist

Here's a business that started as a reasonably successful 900-number information program, only to grow into an even more successful, diversified telephone-based medical and pharmaceutical information service. The company now serves not only consumers, but also physicians, drug stores, health maintenance organizations and pharmaceutical companies. But first, let's talk about how it got started.

Mary Lynn Bell, president and founder of Ask The Pharmacist, Inc., established a 900-number service targeted to pharmaceutical consumers in February 1991, based in Chapel Hill, NC. The telephone number is 1-900-4200 ASK, and the charge is $1.95 per minute. This is a live program staffed by licensed pharmacists.

Prior to launching this business, Mary Lynn was a partner and co-founder of the first statewide employment service that specialized in matching up a pool of licensed pharmacists with drug stores and hospitals that needed temporary fill-in help. Pharmacists also became the basis for her next business, Ask The Pharmacist. While recovering from surgery, with a lot of time on her hands to think about what to do next with her life, her entrepreneurial mind divined a unique opportunity that was emerging as a result of changes taking place in the pharmaceutical industry.

Mary Lynn was aware of the trend that was changing the way over-the-counter and prescription drugs are sold. More and more big drug store chains are beginning to dominate the retail sales, along with significant growth in mail order sales. This has created an environment where consumers are finding it increasingly difficult to receive enough quality information about the prescription drugs they take. This may be one of the reasons explaining why almost half of all prescription drugs are not taken as prescribed. This often results in reduced efficacy, or worse, side effects that need not occur. The days of the small neighborhood pharmacy are gone forever in many parts of the country, but the information needs of pharmaceutical consumers haven't changed at all. Indeed, the complexity of new drugs and multi-drug therapies appears to be increasing, and information about these drugs is often hard to come by.

A licensed pharmacist herself, Mary Lynn Bell understood the needs of her patient-customers, and designed a solution that would address those specific needs: Live advice, exactly when needed, provided by a knowledgeable licensed pharmacist, from the privacy of home. Instead of being forced to ask embarrassing personal questions from the person at the window, who may or may not be the actual pharmacist, while a long line of impatient customers stand behind you, now you can call by telephone at your leisure from home with complete confidentiality.

When Mary Lynn Bell's program got the attention of AT&T, the people there were so excited about the program concept -- being precisely the kind of program they wanted to see using 900 numbers -- AT&T invested a significant amount of money in a national advertising campaign to help promote the program. Here was a program that would serve as an example of the quality professional applications that AT&T wanted to attract into using 900 numbers. This marketing help would have been any IP's dream come true, and obviously AT&T had high expectations for the program. It was not disappointed.

AT&T's investment and a lot of excellent publicity exposure catapulted the company to early success. But that was only the beginning. It wasn't only patients and consumers who heard about Ask The Pharmacist. Drug store chains, pharmaceutical companies and others involved in delivering health care services also took note. Here was a service, already operating successfully, that was tailor made for these companies. They began contacting Mary Lynn about providing live pharmaceutical information services to their customers, by telephone, usually using a toll-free 800 line. Their customers needed this kind of information, and it was easier to use an established service, already operational with well-qualified, telephone-trained registered pharmacists, than to re-invent the wheel by trying to do the same thing in-house.

It became apparent rather quickly to Bell that offering her services to such companies would probably be more successful than the 900 service by itself. After all, it's the same basic information service for the same callers, and therefore totally complementary. The only difference is who pays for the information.

It didn't take long for this side of the business to dominate, and the 900 line, although still operating and generating income, is now playing a minor role in the overall business picture. Because the 800 side of the business is doing so well, demanding all the company's marketing resources, no advertising money is currently spent on the 900 line.

Although the original target audience for the service, the people who actually call the pharmacists, are patient-consumers or their family members, the primary Ask The Pharmacist **clients** are now drug store chains, pharmaceutical companies, health service providers and health care professionals. And not all of them are necessarily part of the health care industry. American Express is a customer, offering toll-free access to Ask The Pharmacist to its Senior Gold Card holders as one of the mix of services provided to its members.

A significant part of the business is now offering custom projects of a specific duration. For example, when a pharmaceutical company launches a new drug or recalls an existing one, it can turn to Ask The Pharmacist to help in fielding questions about the product. In this situation the staff at Ask The Pharmacist become an adjunct to the pharmaceutical company, helping out on an as-needed basis.

The backbone of Ask The Pharmacist is the pharmacist. The pharmacists are the experts in the field of drug information, and the public has a high degree of trust in this group of health care professionals. The staff at Ask The Pharmacist must also be excellent communicators. Each new pharmacist undergoes an intensive in-house training program lasting at least two weeks, bringing them up to speed as good telephone communicators.

Each pharmacist on duty has access to several pharmaceutical databases as well as several volumes of printed reference materials. They also have the capability to tie into the database of the pharmaceutical company they're supporting at the time. Being able to customize the service to the specific needs of the pharmaceutical company client has been critically important.

Mary Lynn Bell is very supportive of her profession. She sits on the Board of Visitors of the Pharmacy School at the University of

North Carolina. Ask The Pharmacist has been recognized by the North Carolina State Board of Pharmacy as a non-traditional practice, which has led to pharmacy students spending part of their formal educational time at the company. Although not required to, Bell has made it her company's policy to employ only licensed pharmacists as telephone consultants.

When asked whether consideration had ever been given to offering recorded interactive services by telephone, Mike Tower, an internal consultant with the company, responded, "the key to providing this type of advice is the ability of the pharmacist to ask a question back to the caller, such as 'what are you being treated for,' or 'what other medications are you taking?' A pharmacist cannot do his or her job without being able to ask the right questions of the patient."

An important reason behind the success of this business is the fact that there is no substitute for live consultation. The printed information that comes with a prescription can be confusing, if not downright frightening. For liability reasons, the pharmaceutical company must list every possible side effect you may experience from taking the drug, regardless of how remote the odds. If you're taking three prescriptions with several pages of warnings you might want the peace of mind that will come from talking to a pharmacist.

When asked why the telephone was such a good medium for delivering this type of service, Tower responded, "almost all of us have access to a telephone no matter what our economic condition. It is simply the most economic way we've found to deliver the highest quality information possible to the largest audience possible."

According to Mary Lynn Bell, "when calling from the privacy of their home, callers are quite forthcoming with a tremendous amount of information, and we're able to help them understand the outcomes of taking their medication properly, and the pitfalls of taking the medication improperly." An indication as to how helpful this program is to its users, there are still plenty of callers to the 900 line despite the fact that there has been virtually no advertising for quite some time. These can only be repeat callers (or their referrals) who saved the articles published about the program when it was launched a few years ago. This in itself is a real testament to the value of this program to its end users.

Chapter 5

Chapter 6
Investment, Finance
& Business Information

Stock and commodity prices are the obvious examples of time-sensitive information that is well-suited to the 900-number delivery medium, and such programs are quite common. The interesting programs, however, are those that offer highly specialized information to a specific group of people.

Until recently, offering business information on 900 numbers was becoming infeasible because most businesses blocked all access to 900 numbers as a matter of policy. Too many employees were running up huge bills calling their astrologers or their "fantasy dates."

AT&T has met this problem head on by instituting a dedicated business exchange for qualified programs that must meet fairly stringent criteria to be assigned the exchange, 900-555. Businesses can now selectively unblock the 900-555 exchange in order to gain access to the wide variety of business information that is available -- or that will be developed in the future.

Limited Partnerships

Limited partnerships aren't considered suitable for conservative investors or for older people who will want cash back within a few years, but for appropriate investors a new 900 application can track and analyze partnership data. You'll be able to learn such things as where the payout is coming from, how the partnership is doing and what your units are worth. Robert A. Stanger & Co., a finance company, has set up 900-786-9600, with a $5 a minute cost. *Aug. 1990 (active)*

Business Info Line Starts
Entrepreneur Magazine Begins a 900 Line

Entrepreneur Magazine has started an Entrepreneur Info Line for those who want to start and operate a successful small business. By calling 900-288-8890, for $2.00 the first minute and $1.00 each minute thereafter, callers can choose business topics from a menu.

The menu includes an entrepreneur's assessment quiz, which suggests the best business to match the caller's personality and skills. Late-breaking business news and small business profiles are also a choice on the menu.

IdealDial of Denver, Colorado designed the service. *May 1991*

Infoline Gets Phone Numbers

Need the telephone number of a business but can't call information because you have no idea where it is located? Dial 900-896-0000 for assistance. American Business Information, in Omaha Neb., is receiving 600 calls a day at a call-cost of $3 for the first three minutes and $1.50 a minute for each subsequent minute.

The company uses a database of 9.2 million U.S. businesses compiled from 5,000 telephone directories. *July 1991 (this was the predecessor of Match-a-Name, below)*

Match-a-Name Update

Earlier this year, when Warren Miller, CEO of Telecompute, a Washington, D.C.-based service bureau, learned of the discontinuance of 900-884-1212 operated by the Gannett Company and *USA Today*, he asked if he might take over the number and they agreed. Miller's bureau had provided the transport for the service, so the transition went smoothly.

Then Miller looked for an operating partner to provide the personnel to operate the database and perform the lookups, and Philadelphia-based National Telephone Enterprises was interested. It leased a database and collaborated on a new name for the service, which is similar to calling directory assistance. Callers to Match-a-Name are provided with a corresponding name and address to a telephone number for $1.49 per minute.

The number, 900-884-1212, was transferred to Telecompute, and application for two additional numbers, 900-555-MATCH and 900-555-4111, was approved by AT&T. Under a joint operating agreement, calls to all three numbers reach a preamble supplied by Telecompute before the call is routed to national operators. Additionally, in July, Telecompute was awarded 555-MATCH by Bellcore, allowing for further expansion in pay-per-call beyond 900 billing.

AN always saw this as an important and useful 900 application. When *Newsday*, the largest suburban newspaper in the U.S., asked us to compile a "900 Sampler" with some of our favorite 900 numbers, we included 900-555-MATCH. Along with *Newsday, AN* received more than 100 calls from consumers wanting to use the service but couldn't due to a typographical error in the newspaper. *Dec. 1994*

Buy Freedom 900 Network

Timed to coincide with Martin Luther King, Jr.'s birthday in January, the Buy Freedom 900 Network (212-575-0876) will have their numbers up and running. Working directly with AT&T equipment, the network plans to operate a premier service business locator and discount hotline. There are many aspects to the network, with some parts of the menu sponsor-driven.

Tony Brown, television talk show host, will introduce the much talked-about Buy Freedom 900 Network across the country. The program is designed to build up the marketing muscle of existing Afro-American businesses and help create thousands of new ones.

In an effort to promote economic empowerment for the Afro-American community, Brown has founded the Buy Freedom Network, a 900 phone system that works a bit like the yellow pages. Callers will be able to dial the number, not yet assigned, to find Afro-American or white-owned companies that sell the products they need. Profits from the calls will go to the Self-Employment Enterprise Fund. This nonprofit group will guarantee loans for minority entrepreneurs who might not otherwise get them. "My dream is to create 50,000 new businesses and take 50,000 families off welfare," said Brown.

The long-time advocate for Afro-American economic empowerment who also writes a syndicated column, believes the potential in the Buy Freedom 900 Network is "awesome."

In explaining how the network will assist the development of new businesses and expand the operations of existing ones, Brown said, "Profits from the call --$1.99 for the first minute and 99 cents for each additional minute -- will be put into a pool. The pool will be the bank from which new and existing businesses can draw capital." *Dec. 1991*

Fax and 900 Applications

Over a year ago, Dynamic Fax started a 900 line coupled with fax that is successful by any measure. At a cost of $3 for the first minute and $2 for each additional minute, callers to 900-329-4982 can connect themselves with the NY and other World Trade Centers. They obtain a fax of available trade leads. These could be for large quantities of foodstuff, real estate or a variety of other trade items. Each item is one page and most callers retrieve 10-15 items per call. Each page takes 30 seconds to transmit.

Dynamic Fax has a customer base of 500 "on and off" users. The line and the six digit file number are marketed through the Los Angeles Journal, the Electronic World Trade Center Network, and the Journal of Commerce.

The company, which has a number of other 900 fax applications, gives callers personal attention and support. Technical assistance requests from new customers, especially foreign speaking, account for at least five calls a day.

The system is uncomplicated. Customers use the fax handset, dial the 900 number, enter the file number, and the fax is immediately delivered.

When *AN* asked about 900 blocking (which other interactive fax companies had indicated was a problem), Dave McKinney, Dynamic Fax president, said, "Callers experiencing that difficulty can call and use a credit card. However, the items are $5." He emphasized that, "Our customers want 900."

Some industry insiders question if the call-blocking devices now available at low cost, which can be used for individual telephones and with great specificity, will open business phones to this type of application. Companies can substitute specific rather than wholesale blocking. *Nov. 1991 (active -- see in-depth profile at end of chapter)*

Journal Finder 900 & Fax

The *Wall Street Journal* is using 900 and fax to provide an instant fax reprint service. The cost is $5 per reprint. Within minutes after calling 900-Find-WSJ and entering a fax number and a three-digit code that identifies the reprint, it is faxed. *Dec. 1991*

Fax & 900 Real Estate Line

Using a database gathered from a variety of sources, General Fax is providing a fax line for real estate investors. For $2 a minute, callers to 900-835-4852 receive on their fax up-to-date National RTC listings, Federal Land Grants, and HUD properties for the New York, New Jersey and Pennsylvania area.

They advertise the number in *Wall Street Journal's* Mart section, local real estate sections and a computer database, Compuserve. Chris Stephano, president of General Fax, said, "Each week we are doing better, we started with eight lines the last week in December, and now we are up to 24." *March 1992*

TRW Credit Report Via 900
IP Seeks Satisfied Customer

For a one-time charge of $28, callers can request a TRW credit report for any business through MTC Information Systems, Novato, Calif.

Cobb Bennett, a spokesperson for MTC, stated that, "The 18-second preamble we use is an important feature. Here is an opportunity for callers to decide whether they want to get the reports and to hear the price again. It legitimizes the effort and as this is a business-to-business application, we want satisfied customers."

MTC purchases the information from TRW and sells it through 900 to the consumer, using Telesphere's phone lines. According to Mr. Bennett, "Callers have the option of receiving the full report by fax, mail or Federal Express."

Clients get the entire contents of the TRW report, which usually averages 3-5 pages. Should there be no report, as is sometimes the case, there is no charge to the client.

Dun & Bradstreet offers two 800-numbers, for credit reports. *Aug. 1990*

Mortgage Rates By Keypad
Do 900 # s Give Lowest Rates?

The Mortgage Bankers Association, which expects phone services that provide mortgage-rate lists to proliferate in the next decade has two words of advice for phone-service users: "Shop around."

The 900 numbers will give callers a computer-generated answer to loan questions by telephone keypad. There are currently 40-50 lines nationally that analyze a caller's monthly income, the price of the desired house and the buyer's monthly debt load, and then give the caller the names of three lenders with lowest rates based on that caller's qualifications. *May 1990*

CD Rates at Tampa Banks Go 900

Residents in the Tampa Bay area can call up the latest rates on Certificates of Deposits (CDs). The CD Rateline 900-Bank-CDS gives CD rates at Tampa banks. Cost is $1.95 for the first minute and 95 cents for each additional minute. *Oct. 1990*

Insurance Ratings Via 900

Businesses and consumers concerned with their insurance companies' financial performance can call 900-420-0400 (now 900-555-BEST) to get their ratings.

This new service -- BestLine -- is an easy way of accessing the A.M. Best Company's rating of the relative financial strength and performance of 3700 insurance companies.

A.M. Best Company, Oldwick, N.J., is to insurance companies what Moodys is to the bond market and has been reporting on insurance companies' operations for 90 years. Callers to the number must have either the A.M. Best company identification number or the National Association of Insurance Commissioners (NAIC) identification number available. These identification numbers can be found in libraries, from state insurance departments and by calling the customer service department of A.M. Best (908-439-2200).

The call is $2.50 per minute. A fax confirmation can be ordered during the call, or a mailed version is available for $10. *March 1991 (active)*

24-Hour Business Hotline Begins

For 95 cents per minute, callers throughout Western North Carolina can get stock market quotations on a 24-hour-a-day basis. The service is produced by The Associated Press in association with member newspapers. *Oct. 1990*

Credit Power Hotline Begins

At the National Consumer League Symposium on 900 Numbers, there was a great deal of discussion related to "rip-off credit card 900 lines." In fact Joselle M. Albrecht, Assistant Attorney General, State of Texas, said she had never heard of one that was legitimate.

Audiotex News informed her that Americom had created "THE CREDIT POWER HOTLINE" to provide detailed consumer credit advice, for $2 per-minute.

Gregory Wood, Americom COO, told *Audiotex News*, "With all the flack over bogus credit and credit card 900 programs, I believe the Hotline will act as a breath of fresh air on a subject that affects everyone."

Perhaps Americom, which has worked to improve the image of 900, will be able to do just that working with Dan Berman, Ph.D. a specialist in consumer credit and a former professor and researcher at the University of California at Berkeley.

Berman is providing practical information on over 70 topics dealing with getting credit, repairing credit and using credit to make money. Free of charge, Americom provides a complete directory of all topics (456 Montgomery St., Suite 900 San Fran., Calif. 94104). *April 1991*

Fax Portfolio Service for $12.50 per Month

Attendees of the Voice '91 Conference saw an exhibition of FAX Interactive, Inc's "FaxTicker." Developed with the Atlanta Journal and Constitution, this is the nation's first fax stock portfolio service enabling subscribers to the service to receive their personalized stock update by the fax. *May 1991*

NOONAN Line & Spot Rates

NOONANLINE, 900-288-5858, provides live, realtime spot rates on the yen, mark, pound and Swiss franc for $2.95 a minute. According to Doug Dressler, executive vice president at Noonan Astley & Pearce (currency brokers), "As far as I know, we are the first to provide spot currency prices. I am excited about the promise of 900. Just starting, we don't have the full picture yet but early result looks good."

This application has three elements that may make it a real winner: 1. repeat callers; 2. callers who stay on the line while they're conducting business on a second line; and 3. a service that fulfills a real need.

Dressler said, "Although there's a menu with 'press one for Yen' etc. and 'five for additional information,' the line is live because of the immediacy of the information." The line, handled by three currency brokers, was developed directly with AT&T."

The decision to go 900 was a practical way of meeting a diversified customer base. When clients need the information, it is an intense need, but in some cases they are unwilling, or it is impractical to pay for a monthly or yearly subscription service," Dressler explained.

One of his concerns was the amount of 900 blocking currently present in brokerage firms. This is a story *AN* will be following to see the impact of blocking or if the NOONAN LINE, in the tradition of Lotus 1-2-3 and Pratt Whitney, can be successful despite blocking. *June 1991*

Comtex & INN Go 900
Shareholders Get Information

Comtex and Investment News Network (INN) announced a joint venture arrangement allowing INN to have Comtex's full-text OmniNews service available through INN's financial 1-900 telephone/fax service.

The problem of closer and effective shareholder communication is becoming increasingly difficult for North American public companies. There is no way of ensuring that non-registered shareholders are informed of pertinent corporate information as more and more companies choose to mail financial information to registered shareholders only.

INN allows public companies the ability to communicate with all shareholders efficiently and effectively through a 900-pay-per-call telephone network. Companies provide information (such as press releases, financial reports) to the INN system. Shareholders then simply use a touch-tone phone or fax to access this information 24 hours a day, seven days a week.

The alliance between Comtex and INN allows INN to provide detailed financial and industry news on more than 18,000 companies, as well as international and national news. *May 1992*

Postal Info HotLines
Rate Info By Phone

Almost -- but not quite -- as prominent in today's office as the telephone is the Pitney Bowes postage machine. Ideal Dial of Denver Colorado, the service bureau, designed and operates two new 900 numbers for Pitney Bowes.

The Postal Information Hotline, 900-896-POST, provides callers with an overview of the recent increase in postal rates and practical tips on how to reduce mailing costs and speed the delivery of mail. The call-cost is $1.50 per minute.

For a $5 call, the Postal Rate Hotline, 900-896-RATE, provides specific details on the new postal rates and offers each caller a detailed, easy-to-read postal rate chart.

The U.S. Postal Service's Postal Answer Line has information on the new rates on their interactive 800 number, 800-624-5225. The menu includes 75 different messages but requires you have the directory (free by mail or in post offices) handy for the message number. *April 1991*

Real Estate Line OnLine
Nine Brokers In Eight States Use 900

The Real Estate Hotline provides listings, open house schedules, relocation information, home buying and selling tips, mortgage rate and more, around the clock with access by local telephone number. Home buyers, working with a menu of options, may choose to connect directly with the listing agent and set up appointments to view the properties. Nine brokers in eight states are using this system, provided by Brite Voice Systems. *May 1991*

Bond Calls Data Via 900
Status Info is Needed

Dow Jones & Co. has taken advantage of the record number of bonds that have been "called" because of low interest rates. One of the biggest problems for bondholders is the lack of information on whether a bond is callable, when it might be called, or whether is has already been called.

For $2 a minute, callers to 900-933-Bond can punch in the bond's CUSIP number and get the bond's maturity date and call date. *July 1992 (active)*

Customs Information

Travel lines have found a home with 900, and one new line focuses on international travelers and importers. Callers to 900-CUSTOMS speak to licensed customhouse brokers.

The line operates from 8:00 a.m. to 5:00 p.m. CST Monday through Friday. Calls cost $5 for the first minute and $3 for each additional minute. The brokers can answer questions about exemptions, allowances, restrictions, duty rates, visa and quota requirements, marking requirements and gifts. Worthington, Ohio-based Scherers Communications is the service bureau. *Dec. 1993*

CNBC Stock Quotes

CNBC, the financial news network, is now offering stock reports via 900. The line, 900-288-CNBC, offers the New York Stock Exchange, the American Stock Exchange and NASDAQ stock quotes.

Calls cost 95 cents per minute. All market data is delayed 15 minutes as required by the financial markets. *Dec. 1992 (active)*

Commodities Line

Speculators looking for advice on buying or selling stocks, bonds, commodities, indices and options can dial the Nimble Traders Stock Market Update at 900-988-TRADE. Calls cost $2.99 for the first minute and $2 for each additional minute.

According to the line's sponsor, MCE Telecommunications, call counts went up with Hilary Clinton's publicized investment of $1,000 in cattle futures. *AN* speculates callers hope to make their own $100,000. *May 1994*

DOC Goes 900
Gov't. News via Fax

Even the United States government realizes the value of 900. The latest economic, financial, and trade news from the Economics and Statistics Administration of the U.S. Department of Commerce is now available in the 900 fax application EBB/FAX™.

The program, which utilizes the FaxFacts Fax-On-Demand System from Copia International Ltd., is part of the Economics and Statistics Administration's Electronic Bulletin Board (EBB).

Users can dial 900-786-2329 from their fax machine's phone and use code numbers to request the most recent updates of government information files. All calls are 65 cents per minute, and callers do not need to be EBB subscribers.

The program works directly out of the Department of Commerce with no service bureau involved. It operates on an eight-line in-house telephone system.

The list of information files is updated each business day. As many as 23 various titles are available, all published by the Bureau of Economic Analysis, Bureau of the Census, International Trade

Administration, Federal Reserve Board, Bureau of Labor Statistics, and U.S. Department of the Treasury. Transfer times vary from two minutes to as long as 21 minutes. *Aug. 1992 (active)*

Public Relations Information
Live Consultant for $3 a Minute

Former deputy press secretary to President Jimmy Carter, Patricia Bario, now heading her own public relations firm in Washington, D.C., has introduced "Do-It-Yourself" public relations using a 900 number.

"Your PR Team," which just got off the ground last month, enables callers to speak directly with a public relations professional, sometimes Bario herself. She acknowledged using the line to promote her new book, "Using Public Relations to Sell Products, Ideas - or You."

"Over the years," says Bario, "we have observed that a number of entrepreneurs ... miss out on the advantages of public relations advice because they think hiring an agency is too expensive. Often they don't need the full service of an agency but simply want to test their own ideas and plans with a PR expert."

In speaking to *AN*, Bario admitted she was surprised at how many free-lance PR professionals were calling. Bario concluded if they used her or a member of her team to bat ideas about for a half hour, even at $3 a minute, they were getting a bargain.

Bario is pleased with her service bureau, D.C.-based Telecompute. For copies of the book, call 900-820-8600 or send $15.95 to Capitol Marketing Solutions, Box 15491, Washington DC 20003. *Feb. 1993*

Earnings Reports by Fax

USA Today on Demand has started faxing complete quarterly earnings reports from selected companies as they're released. The cost is $2 per report. Callers to 900-933-8585 enter the four-digit access code listed after companies in a *USA Today* advertisement.

PR Newswire and Business Wire compile the earnings reports. Call 800-441-5494 for a faxed listing of all available company reports. *Sept. 1992*

Fax & 900 For Corporate Name

Law firms in need of corporate name availability from the Colorado Secretary of State's office can now get that information via automated fax service at 900-555-1515. Firms using the service pay $5 per fax and can make up to three requests at once. The service began in April. Denver-based IdealDial is the service bureau, and all faxes they receive are sent twice daily to the Secretary of State's Office.

A letter to law firms from the Secretary of State's Office helps promote the line. *Aug. 1993*

Racing Form Offers Stock Quotes on 900

Some think stocks are a gamble, but not quite the same as horse betting. Perhaps not. *The Daily Racing Form*, the publication for racing, in response to reader interest, introduced a 900 line for stock quotes information.

Readers call 900-288-5888 for stock information for companies that trade on the New York Stock Exchange, the American Stock Exchange and NASDAQ. Calls cost 89 cents per minute. *May 1993*

Small Business Help-Line
Southwestern Bell Grant Helps 900 Line Start

Small business owners in need of information related to keeping their businesses vital can now turn to the Small Business Help-Line. The National Federation of Independent Business (NFIB) Foundation started the line in late June, aided by a $100,000 grant from Southwestern Bell Telephone.

Callers to 900-820-6342 can hear recorded information on topics such as finance, personnel, marketing, tax matters, starting a small business and legislation and public affairs affecting small business. Calls are 95 cents per minute, and the average call is about four minutes. Callers are given a list of references for further information at the end of each message.

West Interactive is the service bureau for the line. According to William Dennis, senior research fellow for the NFIB Foundation, all proceeds go toward the cost of maintaining it. The Foundation is an affiliate of the Nashville-based NFIB, the 600,000-member small and independent business trade association.

According to Dennis, upcoming promotional efforts for the line will include some direct mail in the five-state region covered by Southwestern Bell. They will also include taglines mentioning the Help-Line on the billing materials sent to customers. *Aug. 1992 (active)*

Stock Analysis

The Dynamic Research Group (DRG) now provides real-time stock analysis and opinion for individual investors through a 900 number. With Denver-based IdealDial as a service bureau, the 900 line uses Vari-A-Bill,, allowing callers to pay only for the options they choose. The 900 number, 900-288-DYNA, has DRG's top-rated stocks, most of the stocks in the S&P 500, as well as other selected companies.

The computerized system recalculates a "comparison score" for each company and generates recommendations to purchase, hold or liquidate. The caller can listen to reports or have them faxed. An 800 number allows callers to purchase information using real-time credit card authorization. *Sept. 1993*

Sprint's 900 SBA On-Line

In an effort to offset skyrocketing costs in providing the long-distance lines into the Small Business Administration's (SBA) computer bulletin board system, Sprint Telemedia is making that system a 900 call for users.

According to Tom Murphy, director of national sales with Sprint Telemedia, when the company agreed to service the bulletin board, called SBA On-Line, Sprint committed $60,000 for the first year. More than 600,000 calls were logged in that first year, and Murphy said Sprint's costs totaled $2.4 million.

SBA On-Line attracts approximately 2,000 calls per day. The switch to 900 reportedly upset some users, with a common complaint being that a call can last 30 minutes or more. Meanwhile, Murphy noted that the 900 charge will merely cover costs for Sprint, not be a moneymaker.

The first minute of connect time is free, then users are charged 30 cents for the next minute and 10 cents for each minute afterward. SBA spokesperson Donna Harper estimated that a 30-minute call would cost users $3.20.

Some services, such as loan and business-development programs and information from other government agencies, will still be available via computer through a toll-free line. *March 1994*

Commodities Research Bureau Line

Knight-Ridder offers its *CRB Blue Line* for commodities investors. Callers to 1-900-454-BLUE get up-to-the-minute commodities prices, trade recommendations and forecasts from the analysts at the Commodity Research Bureau, for $1.33 a minute. *(active)*

Bank Cancels 900 Check Verification

Some banks use 900 for check verification *(AN* July '94). NationsBank Corp also planned to do so beginning July 1, but changed those plans after customer complaints. The proposed service cost merchants and other businesses $1 per call.

A concern of NationsBank was the 900 blocking that many businesses have in effect. It had originally counted on businesses' ability to unblock only the check verification number. However, only large companies using computerized telephone switchboards can unblock at the PBX level.

Wells Fargo Bank of California, First Interstate Bank of California, Society Bank of Ohio and National Bank of Detroit continue to use 900 for check verification. *August 1994*

In-Depth Profile:
Word Trade Center NETWORK

The World Trade Centers Association, headquartered at the World Trade Center in New York, has been offering a pay-per-call fax-on-demand service on 900 and 800 numbers since early 1990. The fax-on-demand service is but one component of its World Trade Center NETWORK, a subscriber computer network linking international traders with some 191 World Trade Centers in more than 140 countries.

One of the most popular features of the WTC NETWORK are the trade leads -- offers to buy or sell goods and services covering virtually every industry imaginable. An importer in Istanbul might post (on the computer bulletin board) his desire to purchase 100,000 pairs of Jockey shorts, and a U.S. exporter can respond directly and close the sale. The fax-on-demand system offers the same information as the computer bulletin board. The only difference is that it is a one-way communication system -- the caller has no way to post his own buy or sell messages. It is also available only to U.S. callers owing to coverage limitations inherent with 800 and 900 numbers.

The telephone numbers are 1-900-FAX-4WTC ($3 the 1st minute, $2 each additional minute) and 1-800-656-1234 ($5 per fax selection, with credit card payment). The reason for offering the 800 number option is to offer the service to those companies or individuals that have a block on all 900-number calls. An informational number describing the system is also available: 1-800-937-8886.

According to George Marx, the manager of the World Trade Center NETWORK, many of the computer network subscribers first used the system via the 900 or 800 line, then decided to upgrade to the full computer network. The fax service is an easy way to introduce people to the WTC NETWORK without having to make any further up-front commitments. Indeed, this is one of the primary purposes for establishing the fax system. Once a caller discovers how useful the information can be, it's likely he will take the next step and go on-line with the WTC NETWORK, which would be more cost-effective for frequent users. Besides trade leads, the computer network offers access to more than 150 databases (such as

Dun&Bradstreet, Moody's, S&P, TRW Credit Reports, etc.), a calendar of WTC events, a business news clipping service, and a currency exchange database.

The service bureau for the fax-on-demand system has been Dynamic Fax, Rockford, IL, since the program was launched in 1990. The program is updated twice a week by Dynamic Fax, by electronically accessing the World Trade Center NETWORK's database of trade leads and downloading the information into its own computer.

The main menu on the fax service offers several choices: a description of the system, summaries of trade offers, specific trade lead details by reference number, and information about the WTCA. A first-time caller without access to any information about the system would first ask for one of the trade summaries. Then, he would identify which trade leads he wanted to follow up on and call the system back with the pertinent reference numbers. He would then receive specific details on both the offer and the company behind it.

On the 900 line the call must be placed from the handset of the caller's fax machine, so that the entire transaction is charged to the caller's phone bill. Further, because this is an interactive fax service with several menu options, it is essential that the fax machine be equipped with a touch tone keypad. On the other hand, with the 800 line, the caller can input the destination fax number he wants the information directed to.

When asked who were the callers to the fax service, Marx indicated that, "they are primarily small to medium sized traders of some kind -- importers, exporters, agents." Further, they tend to be fairly savvy in international trade, as the purpose of the program is to facilitate worldwide trade. Marx also knows that many of the callers have been long-term customers of the service because he will get telephone calls from users who tell him they have been using the service for years. And some of these callers have apparently never felt the need to upgrade to the computer network -- perhaps they are only occasional users or they don't own computers.

The WTC NETWORK fax service is marketed through a variety of business and commerce publications in the U.S. One of the most important publications is the *Journal of Commerce*, published by

Knight-Ridder, which includes the WTC NETWORK trade leads biweekly. Some of the other publications used for advertising are *Foreign Trade, U.S. Latin Trade, Free Trade News, World Trade, International Business, Business America*, and several others targeted to international traders. In many cases the WTC NETWORK or some of the other U.S. Trade Centers will enter into a mutually beneficial relationship with a publication, providing valuable trade lead data that will benefit the publication's readers in exchange for free exposure or better advertising rates in promoting the fax service or the WTC NETWORK. This often takes the form of an abbreviated trade lead that can be published directly within the relevant publication, then referring the readers to the 900 & 800 numbers for more information.

Call volume has been "growing at a pretty good pace until the beginning of last year (1993)," says Marx, "but then it levelled off or even dropped a little." He has no idea of why this happened, other than the fact that there are now more sources for getting trade lead information out there, so perhaps increasing competition is eating into the call volume. Nonetheless, the fax service itself has been a financial success from the very beginning. In fact, "we couldn't believe the success it was when we first offered it -- it was a plus from day one," continues Marx. As a percentage of the total revenue generated by the WTC NETWORK as a whole, however, the fax service is a fairly small, albeit positive, contributor.

We decided to give this system a try to see how it works. After dialing the 900 number, we selected the "trade offer summaries" option from the main menu. We were then given a sub-menu with the following choices: Offers to buy, offers to sell, and other trade opportunities. We chose offers to buy, because maybe someone out there would be interested in buying a million copies of this book!

It took a total of about 5 minutes, or about $11, to get a 5-page list of trade leads in 12 categories: Non-processed Animal and Vegetable Products; Foodstuff; Chemical and Mineral Products; Plastics, Rubber, Wood and Building Materials; Textiles; Precious Metals, Jewelry; Electronic and Mechanical Appliances; Vehicles, Aircraft and Ships; Medical, Surgical and Health Care Products; Manufactured Articles; Miscellaneous Other; and Finance, Real

Estate. Unfortunately, the fax machine was set on "fine," and it would have taken less time had it been set at a lower resolution, saving a few dollars.

There were a total of more than 200 trade leads, and unfortunately, no offers to buy books. We tried to find something interesting and ended up choosing "Used Levi's 501 and Vintage Clothes" from the Textiles category. We redialed the 900 number and entered the 6-digit reference number that correlated with that lead. This time it took only 2 minutes to get a one-page description of the offer.

This trade lead originated from the World Trade Center in Bangkok, Thailand and read in part as follows:

Offer to Buy: Used Levi's 501 and Vintage Clothes
Used Levi's 501 (Made in U.S.A.) only blue colour min.
order 1500 pairs per month.
Vintage Levi's, Wrangler, Lee, Maverick - Denim Shirt,
Jacket, and Over-all
Reliable Suppliers are Needed. We Buy Cash.
Pls. Contact Us: M.I.T. Co., Ltd. Fax: 66-2-332-6611

It is easy to see how this system can be very helpful to experienced international traders, or anyone else with something to buy or sell internationally. The fact that the program has a high percentage of loyal repeat callers attests to its usefulness. And the idea of using a fax service to introduce people to a computer network service is both imaginative and quite promising for other applications as well.

Chapter 6

Chapter 7
Sports

Many major newspapers such as *USA Today* have their own 900-number sports lines, which are published in the sports section of the paper. This is a way of providing an additional service to the reader, allowing him or her to be able to get the latest scores that didn't make the paper.

Such programs are naturals for newspapers, complementing rather than competing with the printed information, which by its nature cannot be as timely. Newspapers are in the business of providing information, not selling paper, and a 900 number is simply one more tool to be used to deliver the information.

Penn State Line Begins
ICI Uses Direct Marketing to Promote Line

Service bureaus and IPs are turning to direct mail as a way of marketing their 900 applications. Christine Clark, at Interactive Communications Concepts (ICC), said, "With this current 900 promotion, the print and broadcast has helped focus the caller. We attribute the program's strength to direct mail."

The direct mail piece, produced by the Pittsburgh-based company, went to 70,000 Penn State Alumni. Called the *All-Time Greats Gazette*, it is a four-page blue and white tabloid promoting "All-Time Great" voting.

For $1.25 a minute, Lion Line Football (900-535-0003) offers fans the opportunity to vote for the Penn State "Football All-Time Greats" along with complete Penn State football pre- and post-game information.

The Gazette features the talent callers hear on the line. In addition to the talent, the Lion Line features brief, biographical sketches of the player candidates each week, so callers can refresh their memory. They are able to find previous weeks' winners as well.

The 12-week program began on September 1 and will run through December 12. This is followed by an election finale which includes voting for the all-time greatest offensive player, all-time greatest defensive player and all-time greatest overall player. If callers vote correctly for all three winners, they could be winners of a grand prize package.

The All-Time Greats Gazette makes all of this clear, and readers learn exactly how to play. Prizes are described and other offers, such as opportunities to win a trip to the "Lion Line Football Bowl" or reserve a free all-time greats collectors edition poster, are featured. *The Gazette* has a great look and should attract repeat callers. Lets hope it catches the eye of the alums, and they call and call.

For *AN* readers who are Penn State fans, the three players named as Defensive Tackles for the All-time Greatest Team are Mike Reid, Bruce Clark and Matt Millen. *Nov. 1991*

McMahon Sports Line
QB's Outrageous Opinions Offered

Celebrity lines continue to multiply, and pro football's original bad boy, Jim McMahon, will be presenting his controversial opinions on -- well simply everything -- for $2 per minute on 900-988-6229.

Bringing this new application to the public is GV Communications Corporation, a wholly owned subsidary of Regal Communications of Fort Washington, Penn. The Chairman of the Board of Regal Communications is Arthur Toll, the president of the Gateway Group. *Nov. 1990*

W. Payton Goes 900
Street Smarts Game For Scholarships

Look for positive press for 900 coming from a former Chicago Bear running back. Future Hall of Famer, Walter "Sweets" Payton -- and he is, developed a Street Smarts Game with an anti-drug message. Proceeds from the game are going to the Halas/ Payton Foundation to be used for scholarship funds. *Nov. 1990*

Lacrosse Scores With 900
Rotary Phones Not a Problem for Fans

The Lacrosse Scorline is the first 900 number to report on that sport. The Lacrosse Foundation, in conjunction with Lacrosse Magazine and Advanced Telecom Services, gives fans updated men's and women's Division 1 and Division 3 scores. Collegiate polls, major indoor Lacrosse League results and standings are available as well as feature news and interviews.

Updated twice weekly, the Lacrosse Scorline (900-454-SCOR) costs 99 cents per minute. Partial proceeds are donated to The Lacrosse Foundation. The interactive program uses ATS VoiceTone for the voice recognition of rotary dial phones. *May 1991 (active)*

Baseball Fantasy League Challenge
Player Statistics Count for Prizes

According to Dave Bresnahan, the creator of THE BASEBALL FANTASY LEAGUE CHALLENGE, it's an exciting way for baseball fans to keep track of their favorite players' stats and also have a chance to win cash prizes.

Callers to 900-288-7979 choose a starting line-up including nine players, one for each position and as a tie-breaker, three outfielders from the major league roster. Next the participant enters the number of points they think their selected team will score in the upcoming week. The team they select is entered for one week.

Winners are listed in the display ad of *Baseball America* and *USA Today Baseball Weekly*. The top three weekly winners receive $500, $250 and $100 prizes. There is a $5000 end of the season prize for the highest one week total.

Call-cost is $1.99 per minute. IdealDial of Denver Colorado, is the service bureau. *May 1991*

Football Results

Call the Football Results Line at 1-900-420-5888, and for 75 cents per minute get not only the latest scores, but also the latest Las Vegas betting lines, stadium weather forecasts, and *College and Pro Football Newsweekly's* winning picks. *(active)*

Serious Armchair Bettors

Bettors can call for instant racing results at the Meadowlands and Monmouth Race Tracks in New Jersey by dialing 1-900-990-2800. *(active)*

Other serious bettors can call *Pete's Picks* at 1-900-776-5353 *(active)* for $2 per minute for the latest picks in numerous college and professional sports, or for $10 per call, they can call *Billy's Power Parlays* at 1-900-454-4667.

Computerized Phone Golf
Teleline and Flair Offer Simulation

The Ballantine's Challenge offers a computerized golf simulation that includes course dimensions and hazards, in conjunction with realistic sound effects. Teleline Inc. and Flair Communications have created a 900 number to select club requirements and swing strength for each shot. By calling 900-288-GOLF, a computer will generate the results and keep track of scores. Additional information is given for each shot.

Each golfer who completes 18 holes will be eligible for prizes. The charge for this service is $1.10 per minute. *July 1991*

Indianapolis 500 Pit-Talk
Callers Hear All

When the green flag waved to mark the start of the Indianapolis 500 this past May, 10,000 fans called 800-432-Talk and punched in their credit card numbers to hear the live conversations of the drivers and their pit crews.

The number was advertised on ABC during the race. Callers paid $1.50 per minute, and the average call lasted 5.8 minutes. Call Interactive, the service bureau, assisted Northern Lights Communications of Minneapolis, and proved that the technology was more than up to the job.

Now that this has been done Jason Gould, the director of the broadcasting programming group, looks forward to next year when to help offset the considerable advertising costs, he plans to have the program sponsor driven. *July 1991*

Teleline Aids Olympic Spirit

Last month during the Winter Olympics, callers to 900-884-SPIRIT could leave a message for American athletes.

Teleline, the service bureau handling the promotion for Maxwell House coffee, created a system in which all 189 Olympic athletes had a "voice mailbox." Teleline downloaded the messages onto audio cassettes, and shipped them to the Olympic Village in France.

The Olympic Spirit telepromotion was kicked off with a 50 million FSI (free-standing insert) newspaper drop on January 19. Additional promotion was provided through in-store displays and print magazine advertising.

The cost of the call was $1.50 per minute with an average length of two minutes. The promotion included a sweepstakes overlay with over 3,000 prizes. *March 1992*

900-FAN-A-TIC

With college and high school football a way of life in the South, it's widely known by college football fans who the hot high school prospects are. Recruiting information for fanatical fans is nothing short of religion, every fan hoping they'll have the data to recognize the next Joe Namath.

Alabama "Fan-A-Tics" are able to access some of the most knowledgeable high school football analysts by calling their 900 number. Jeff Whitaker, Forrest Davies, Max Howell and former University of Hawaii quarterback Jeff Duva make their analyses of the recruiting scene available to callers.

Max Howell, on 900-407-0000 [great number isn't it?], received 470 calls at $1.49 a minute to his recorded message -- on a slow day. *April 1993*

Bass Tournament News on 900

The Bass Anglers Sportsman Society now offers 900-990-BASS through which bass fishing enthusiasts can hear daily results from professional tournaments.

For $1.95 for the first minute and 95 cents for each additional minute, callers can hear tournament results, interviews with top finishers and tips from professional fishermen. The average call length is three minutes. *Jan. 1994 (active)*

Jets Fax

Jets fans will get up-to-date information on the team through Jets Fax from Phone Programs, Inc. A one-page fax is sent to the subscriber's home or office every business day at 4 p.m. Eastern Time. Faxes include late-breaking news, injury updates, transactions and statistics.

The service will run from August 29 through December 23, and will cost $59.99 payable by check or money order. *July 1994*

Burns' Football Analysis on 900

Jerry Burns, former head coach of the Minnesota Vikings, is featured on a new 900 line. Callers to 900-BURNSIE can hear Burns talk about and analyze recent Vikings games. Calls cost $1.49 per minute. *Oct. 1993*

SportsFixes on 900

Michael Lubell, president of SportsFix, a new information provider, started advertising the beginning of last month and said, "Although our call count has a ways to go, I am very pleased with our staff of experts and the product we are offering."

The live application offers skiers, golfers and tennis players advice on sharpening their skills. The $1.95 per minute line at 900-896-PROS draws from a database of customized tips and equipment evaluation. Live operators discuss proper equipment, playing strategy and improvement. Calls average seven minutes.

The line also offers faxes of tips or other information through 800-935-PROS, plus a video review service. Players can have their videotaped action reviewed for a $19.95 charge.

Callers can also become SportsFix™ charter members and receive 50 minutes of calls for $60, product coupons, free t-shirt and reduced price on video evaluation. *May 1993*

Sports Lines

Sports continues to be strong on 900. Two recent lines feature news on the sports teams of St. John's University and baseball playing tips from a member of the Houston Astros.

St. John's fans can call 900-407-SJU1 and pay 99 cents per minute for news about the university's sports teams. The St. John's Sports Report Line began late last year and operates year round.

John Massarelli, Houston Astros catcher, gives baseball hitting tips to callers to 900-454-HITS. Calls to the Baseball Hitting Hotline cost $1.49 per minute.

Advanced Telecom Services Inc. is the service bureau for each of these lines. *May 1993*

In-Depth Profile:
Phone Programs, Inc. -- Sports Phone

This is an information provider that was around even before the days of pay-per-call's beginnings. Chairman, CEO and co-founder Bruce J. Fogel was there from the beginning in the early 1970s, when Phone Programs began offering sports scores and information to New Yorkers on regular phone lines. So were co-founders Fred Weiner, Vice Chairman & COO; and Mark Goldman, Vice Chairman & CFO.

Phone Programs got off to a fast start, and over the course of the first 18 months of operations call volume was built up to some 3 million calls per month. At this time the program was supported by advertisers -- recorded messages such as, "these scores are brought to you by....." Despite the impressive call volume, "It was a case of the operation being a success but the patient died," says Fogel. "We could not sustain advertiser sales because New York Telephone couldn't or wouldn't tell us whether it was one person calling 3 million times or 3 million people each calling once, and we knew it was neither."

In 1974, when Phone Programs gave N.Y. Telephone the news that it would have to pull the plug on the program, costing N.Y. Telephone 3 million calls worth of monthly revenue, the phone company decided to contract with Phone Programs to continue providing the service as the producer of the Sports Phone program, as it had now become named. New York Telephone paid Phone Programs a monthly fee to continue operating the program.

In those early days there was no revenue "sharing" between the phone company and the information provider. Rather, the phone company contracted with an information provider to supply the desired information to the public.

This proved to be an attractive service for other local telephone companies too, and by 1981 Phone Programs was serving 20 cities plus AT&T's long distance 900 service (1-900-976-1313), which was not yet a premium-billed pay-per-call service. The cities served included Tampa, Minneapolis, Cleveland, Chicago, Detroit, and Buffalo, to name a few. It was in Chicago and Detroit, however, where Phone Programs concentrated its efforts, and after New York, those two cities continue to be the largest markets for the Sports Phone program. During this period the 976-1313 number became widely known as *the* sports information number.

Up until 1982 Phone Programs was compensated by the telephone companies as a service provider, and the benefit for the phone companies was basically twofold: providing a service to their customers while increasing telephone usage and revenue. When AT&T was forced to break up in 1982, one of the new restrictions placed upon all of the resulting telephone companies (a smaller AT&T, the 7 Baby Bells, plus independents and new companies) was that they could not be in the business of providing or selling information. This meant that the information content had to be separated out from the transport element, because the phone companies could still obviously earn money from transporting information over the phone lines, they just couldn't collect for the information content.

This created the need to restructure the whole payment mechanism so that the information provider and the phone company shared the revenue in an equitable manner consistent with the service provided. Hence premium-billed pay-per-call was born on 976 lines, where an independent information provider was able to earn money by selling information. Fogel calls this "revenue sharing" with the telephone company, which indeed it is. After the AT&T breakup, not all the phone companies decided to stay in the 976 business. "As of January 1st, 1983 Illinois, Michigan, New York, Pennsylvania, New Jersey and AT&T stayed in the 976 information business by revenue sharing with the information providers," says Fogel. "Everyone else dropped out of the business or was not as yet in the business." This is one of the primary reasons why Phone Programs is big in New York, Chicago, Detroit and with AT&T 900 long distance: they didn't drop out of 976 back in 1983.

The local 976 numbers in each market were each customized to those markets by leading off with news and scores on the local teams and sports. It should also be noted that from the beginning to this day, due to equipment limitations, the New York 976 service has always been a simple one-minute recorded program that is updated almost continuously. The call charge is a flat 40 cents for one minute. The national 900 program, on the other hand, has no such limitation and the recorded message can run as long as 5 minutes, depending upon the season and what is going on in the sports world at the time. The charge for the national program (1-900-976-1313) is 95 cents per minute.

Up until 1990 the programs for all 20 plus cities were all updated and broadcast from facilities in Detroit, Chicago and New York. Since 1990, all the programs are broadcast from one facility in Elmont, New York, where all the equipment is located. Phone Programs now has some call-handling equipment in place in some of its other markets, but the main headquarters is in Elmont, which looks much like a very big radio station.

This facility now employs some 70 people, in about 20 thousand square feet of office and studio space, and has 5 satellite dishes on the roof. Phone Programs also uses some 200 sports stringers around the country to feed in the scores and other news continuously. Phone

Programs now offers several different sports programs, including some which use live announcers who are calling the game as they watch TV (hence the satellite monitoring). In fact, most of the people who work there are sports broadcasters for both the live and the recorded programs.

Many of the programs today are custom designed and private labelled for Phone Programs clients, such as *Newsday* and *Sports Channel*. In most cases the preamble is changed so that the caller never knows that he is reaching anything other than that media company's own in-house sports line.

The actual frequency of the program updates depend upon what is going on at the time. For example, during football season on a Saturday afternoon there may be 90 NCAA games to cover, and the total message (for the national 900 program) could run for up to 4 minutes, so by necessity the updates cannot be more frequent than 4 minutes. On the other hand, on Sunday there may be only 8 NFL games, and the message may be only 2 minutes on the same program, and the updates would be every 2 minutes. According to Fogel, they are the fastest results program available across all sports categories. Some competitors might match them in one or two sports, but no one comes close in all the different sports categories.

Phone Programs has always been first and foremost an information provider. The fact that it was started by people with heavy marketing backgrounds helped a great deal in its eventual success. According to Fogel, "We don't know ____ about telephony. Our expertise is marketing." Except for the private labelling for media company clients, Phone Programs does all the advertising of its various programs all over the country, using a wide mix of advertising techniques and media.

Phone Programs also creates ancillary products, like league and team hotlines, such as the *NBA* and the *Rangers Hotline* (also the Knicks, the Islanders, the Jets, the Giants, etc.), that can be used to help generate exposure for its other programs. This is done on a "cash and trade basis" with each individual team or franchise, according to Fogel, "in exchange for advertising the Sports Phone product in their yearbooks, in their programs, on their screens in their arenas and in their stadiums."

Why do people continue to call Sports Phone when there are so many sources of free sports information? According to the research conducted by Phone Programs, "people want instant information when they want it," says Fogel. "If you want to know on draft day who your team drafted, or if you're an alumnus and want to know a score, and you're really a sports nut, you'll pick up the phone to get that information, and you'll pay for it."

Although Phone Programs used to be its own national service bureau with all the necessary call-handling equipment, it now uses West Interactive as its service bureau for the national programs. "We decided we didn't want to be in the technology business," says Fogel. The company now concentrates on providing a quality information product, and on the marketing necessary to convince sports fans to use their product.

Phone Programs can take much of the credit for getting AT&T seriously into the 900 business. In 1983 Sports Phone was getting only 5 cents per minute from AT&T, but was still able to get by because overhead was so low. "In 1986 and 1987 we went to them with two events: One was WrestleMania I and the other was the Hagler-Hearns fight," continues Fogel. Each event was covered live as if by a ringside announcer, and each one generated hundreds of thousands of calls. Phone Programs did this as a test and to prove to AT&T that there was potential in this business, but only if information providers could earn more than a nickel per minute. Although they lost money on this test, it achieved its purpose in convincing AT&T to take 900 seriously as a vehicle for IPs of all kinds to sell information.

Largely as a result of these events, plus the behind-the-scenes work of Corey Eisner and Bruce Fogel at Phone Programs, shortly thereafter AT&T instituted "sponsor selected" pricing. This allowed the IPs to set the price of the call, within certain limits, and the telephone company collected the entire bill and paid out the IP's share after deducting the transport charges and collection fees. Thus was born the current version of premium-billed pay-per-call services, and the industry began growing rapidly as IPs lined up to sell information over the telephone.

Phone Programs uses all types of advertising media in promoting its programs, but by far the most effective has been television advertising during sports events. Cable TV is now used extensively because it gives Phone Programs the ability to "narrowcast" the commercials during specific sporting events. As long as the call volume is up there, as it has been with Sports Phone, "you're in the enviable position of being able to pay for the waste circulation that television offers," says Fogel. And fortunately, the sports information programs have never suffered from a significant chargeback problem.

Now that they have been in business for so long, having experimented with all kinds of advertising strategies and media, Phone Programs is pretty confident about what advertising works for them and doesn't feel the need to continue with a lot of testing and measurement of various ads. The advertising budget for all the information programs (besides sports, it also offers lottery results and weather programs) is well into the 7 figure range annually.

When asked where he felt the future of the Sports Phone would be, Fogel responds,"I'm not one who believes the 500-channel information superhighway is around the corner." Fogel expects Phone Programs to continue to be an information provider into the foreseeable future, but the delivery medium will change, and he expects to be delivering this information on-line at some point in the future, when there are enough modems and computers out there to support the business.

"If you were to ask 'Will there always be people who want instant information?' the answer is yes," states Fogel. "Then if the question is asked 'Will the telephone be the only way to get our information in the future?' the answer is no. The primary delivery mechanism today is the telephone, but that will not be the case in 10 or 15 years. We are not married to that $20 dollar box." The bottom line is that Phone Programs is very good at what it does best: pulling together the information and marketing its services to sports fans. The actual means of delivery is clearly incidental, and the company can use whatever methods are both available and effective at the time. That could be a "computerphone" or some device we can't even yet imagine.

Chapter 8
Environmental Information

Environmental information is a category that has two of the most important components of a successful 900-number service: time-sensitive information and specialized information. Different people need different kinds of environmental information for widely different reasons: travel, health, safety, economics or recreation. Whether it's air pollution levels or ski conditions, environmental information services will continue to be an important application for the pay-per-call industry.

Surfers Wait To Report

Picture this -- public phone booths stretched across the hot Texas coastline, the sun just beginning to rise from Galveston to South Padre Island. Standing next to the phone booths, veteran surfers anxiously waiting for the phones to ring.

Once they ring, these surfers report into surf central. Surfers who punch 900 USA Surf, enter an area code for a report that allows them to pinpoint the best wave action. *Aug. 1990*

Budweiser Promotes 900 Ski Line
Portion of Proceeds to Not-For-Profit

As part of a Budweiser promotion, skiers around the nation can now get daily slope conditions and a chance to win a cameo role in a skiing film being shot in New Zealand, or the grand prize of a skiing trip for two to New Zealand.

Through April, by dialing 900-896-2-BUD, users get weather and slope conditions in eight regions, updated twice a day. Callers can stay on the line for specific ski condition reports from more than 483 ski areas.

Each call is 95 cents per minute with portions of the proceeds to be donated to Global ReLeaf, an international environmental citizen's action reforestation program. *Feb. 1991*

Competition for Ski Lines

Now 900 ski lines have additional competition from the Weather Channel. After dialing 900-WEATHER, callers can cut right to the ski news by pressing touch-tone three and then two.

The cost of the call is $.95 cents per minute and gives ski conditions at 400 resorts in the U.S. and Europe. *March 1992*

Will Weather Delay Santa?

Although Santa lines in the past have given the industry a black eye -- this one seems safe. Holiday worriers can check the weather at the North Pole by dialing 900-Weather. Starting in December and continuing through New Year's, the American Express weather telephone service will offer weekly updates for the North Pole, in addition to their usual 600 other locations in the world. The call costs 75 cents a minute, has a seasonal message, and changes weekly. *Dec. 1990 - Jan. 1991*

900 Driving Report Line

A 900 application inspired by winter weather is the Washington State Department of Transportation's Mountain Pass Reports at 900-407-PASS. The line reports on latest weather and road conditions for 35 cents per minute. The service covers precipitation, any accumulation on roadways, temperature, and whether traction devices are advised or required.

Callers unable to access the 900 line can call a toll and 800 number for the information. *Feb. 1994*

Any Port in a Storm

Boat owners now dial a 900 number before setting sail. Weather Watch, at 900-933-BOAT, gives official National Oceanic Atmospheric Assoc. weather radio broadcasts for any port in the U.S. Calls cost 98 cents per minute, and callers need to know the area code of the port. The Boat Owners Assoc. of the U.S. sponsors the line. *Mar. 1993 (active -- see in-depth profile at end of chapter)*

Dialing for Willard and the Weather

The NBC Weatherline at 900-WILLARD offers forecasts for more than 850 cities, at 95 cents per minute. It is available across the country, except in Louisiana, a state which does not permit 900 calls. Willard Scott, the *Today* show weatherman, greets and instructs callers, who can hear current weather conditions and three and seven-day forecasts in an average two-minute call. NBC News is donating the proceeds to charity. *Nov. 1992*

Hurricane Andrew Hotline
Provides Storm Info

The number of calls to Accu-Weather, Inc.'s Hurricane Hotline at least doubled and tripled as Hurricane Andrew approached the United States in late August, said Joseph Sobel, Accu-Weather senior vice president.

"In a storm situation there's new updated information every couple of hours," Sobel added. The hotline, which operates through individual 900 numbers for participating media partners, provides National Hurricane Center advisories and Accu-Weather's commentary and forecasts. Frequent updating begins when a storm is due to hit land within 12 to 24 hours.

More than 200 newspaper, radio and television stations are media partners for Accu-Weather's 900 weather services, including the Hurricane Hotline. The hotline is in its second year and remains in operation from June 1 through November 30. The cost to callers is 95 cents per minute. Each media partner typically receives 30 cents per minute for each call, the average of which is two minutes. *Oct. 1992*

Hurricane Hotline Returns

For the third year, Accu-Weather, Inc. is offering the Hurricane Hotline to media partners. The hotline provides callers with National Hurricane Center advisories and Accu-Weather commentary and forecasts.

The line operates from June to November. Callers pay 95 cents per minute. Media partners typically receive 30 cents per call. Accu-Weather currently has more than 200 media partners, each with their own 900 number.

The hotline is updated more frequently in a storm situation (*AN* Oct. '92). Last August, calls at least doubled or tripled as Hurricane Andrew approached the United States. *Sept. 1993*

New Ski Line

In tune with this winter's heavy snows, the National Outdoor Network has launched a snow hotline for Michigan and Wisconsin at 900-988-9086.

At $2 for the first minute and $1.50 for each additional minute, the line gives callers current weather, road and trail conditions, local restaurant and lodging information, as well as weather forecasts. Callers can access info for the entire state of Michigan or Wisconsin or just specific locations. Covering the entire state takes about five minutes.

The line began Dec. 1 and will remain operational until March 20. For information, write the Network, P.O. Box 178, Wellston, MI 49689. *Feb. 1993*

In-Depth Profile:
BOAT/U.S. Weather Watch
Instant Nautical Weather from N.O.A.A. for Boaters

Established in January 1992, this 900 line is targeted to recreational boaters across the U.S. who want instant access to the detailed marine weather information that is continually broadcast by the National Oceanographic and Atmospheric Administration (NOAA) and the National Weather Service. The telephone number is 1-900-933-BOAT, and the charge is 98 cents per minute.

The line is sponsored by The Boat Owners Association of the U.S., widely known as BOAT/U.S. for short. According to Dave Pilvelait, Director of Media and Community Relations at BOAT/U.S., based in Alexandria, Virginia, it is the largest organization of recreational boat owners in the country with just over 500,000 members.

This is a real-time weather program. After calling in to the main program, the caller selects, by telephone area code, the location of the NOAA weather station from which he or she wishes to hear the latest marine weather conditions. There are about 70 NOAA weather stations to choose from, and the caller must know the area code before calling. Most stations are along the coasts and the Great Lakes. After making the selection, the caller is connected to the live broadcast from that station, hearing the exact same broadcast that is transmitted by radio.

The service bureau is INCOMM in Atlanta, which serves as a switching station, taking all the incoming calls and then forwarding them the appropriate NOAA station, by regular telephone line, terminating at a device located at each station that is loaded with a tape of the latest weather broadcast (the same tape that is broadcast live over the air).

The program was developed by The Weather Radio Network in Nashville under contract with NOAA, and is also used for other 900-number weather programs that are not marine weather related. According to Pilvelait, The Weather Radio Network has some $500,000 invested in the overall system, including equipment and software. The BOAT/U.S. Weather Watch line simply uses only those NOAA stations that specialize in broadcasting marine weather.

The average call is three minutes, and most people can get the information they need in that time without listening to the entire tape, which is typically seven minutes long. There is always a measurable increase in call volume when foul weather is imminent or when a storm is brewing along the coast or in the Great Lakes.

This is the only system in effect that makes NOAA weather information available by telephone. Prior to the availability of this line, boaters could hear the broadcast by VHF radio only, usually from their boats, hearing only the local NOAA station, which typically has a broadcast range of about 40 miles. According to Dave Pilvelait, "this service is attractive to boaters who will be cruising beyond the coverage area of their local NOAA station." Someone planning a trip down the east coast along the Intracoastal Waterway, for example, could call in advance to ascertain the weather conditions at points along the way.

Not every boater has a VHF radio, particularly at home, and a prudent boater planning a long cruise will want to plan ahead. Pilvelait says that most of the calls to the service occur on Thursday evenings and Friday mornings, obviously from boaters planning a weekend trip.

Why is the NOAA weather better than the local TV weather broadcast, and why would anyone pay for it? Because it is very specific marine weather that is not otherwise available, with tide, current, wind and other specialized information that is of interest only to boaters. As all experienced boaters know, the expected weather conditions rank up there among the most important things to know, second only to knowing where you are located (otherwise known as navigation!).

Another feature of this program is the ability to patch directly into the storm advisories transmitted by the National Hurricane Center in Coral Gables, Florida. This information is available during the hurricane season on both coasts.

The program is marketed through several channels. The primary advertising vehicle is the bi-monthly membership newsletter, which reaches some 500,000 association members. Space advertising in regional boating magazines is another marketing avenue that has been used with some success. BOAT/U.S. has also been quite aggressive with its efforts in generating publicity, getting good exposure with news releases and public service announcements.

Pilvelait said that they had tried spot display advertising, but it was quite expensive and did not prove to be a viable option. The only paid print advertising that met with any degree of success was through sponsoring a regular segment in the regional boating publications, such as a tide chart, with a banner showing the 900 number, which appeared in the same section of the magazine every month. The readers would come to know where the ad was located and could find the number easily. "Consistency is the key" said Pilvelait.

Radio broadcast advertising was tested, using both 15 and 30 second spots, and the response was quite poor. Cable TV was no better. They are now testing some direct mail marketing options, including a card deck that mails to recreational boaters in Florida.

BOAT/U.S. recently produced a Weather Watch pocket guide that lists all the area codes and all the NOAA stations with instructions on how to access the service. This pocket guide is marketed through press releases, distributed at boat shows, and mailed to members upon renewal of their membership. It's a wallet-size heavy duty plastic card that can be kept as a permanent and convenient reference.

Call counts have been gradually increasing, and so far the only effective paid advertising has been the long-term sponsored program tie-ins, such as the tide tables in the regional boating publications, and that has been "only marginally successful," according to Pilvelait. So far the program has not been profitable in the sense that it has not yet generated income in excess of the advertising expenditures.

Obviously, testing different advertising methods is expensive, and this program is still trying to find the most cost-effective ways to generate calls. Pilvelait feels that they need to go more directly at their target market, which is one of the reasons for developing the pocket guide and for trying more direct mail. For example, BOAT/U.S. offers an Affinity credit card program to its members, and a message about the program is now printed on the outgoing bills.

Another effective promotion has been to offer demonstrations at boat shows. BOAT/U.S. also conducted a drawing, through the newsletter, entering anyone who asked for the pocket guide in a drawing for an hour of free time on the program. It received a "couple thousand" requests for the pocket guide as a result , and this proved to be an effective way to raise awareness about the program among the membership.

BOAT/U.S. has just begun offering the service on an 800 number with a prepaid calling-card type option, at a reduced rate of 85 cents per minute for several different blocks of time (minimum 15 minutes). The latest version of the pocket guide in fact has a PIN number printed on it, along with instructions on how to activate that PIN number, and callers are instructed how to purchase a block of time which is paid for by credit card during the call, all accomplished interactively with the service bureau's equipment.

Offering the service on 800 gives a price break to the heavy users of the program, and is more widely accessible than 900. Some potential users may have 900 numbers blocked at their home or office, and many boaters are now using cellular phones, which cannot currently access 900 numbers.

The revenues generated by the program are slowly increasing, and the breakeven point will probably be reached sometime in 1995. However, the fact that the program has not been a huge money-maker for the association is not really an issue. The purpose for the program is to better serve the membership, and this program is simply one of many services provided for the benefit of the members and the boating public.

Chapter 9
Lifestyle,
Travel & Leisure

This category, if indeed it can be called that, is admittedly fairly diverse, particularly with the inclusion of the "Lifestyle" subject matter. What we tried to achieve here is to segregate subjects that deal with the personal lives of people, as distinct from the other subject categories in this book.

900 Number Reaches Mom and Dad

The recorded voice of "Mom" or "Dad" will offer advice on seven different topics -- loneliness, anger, the blues, your personal life, career, health and how to deal with holiday doldrums. This Telesphere line is targeted for the 25-45-year-old who is career-oriented, works long hours and lacks intimate friendships. To access the parents, call 900-456-HOME for a charge of $2 for the first minute, 95 cents for each additional minute. *Feb. 1991*

Talking Ads

The *Atlanta Journal and Constitution* is leading the revolution in marketing technology. The newspaper's Personals section offers an optional voice mail feature. For an additional $10 anyone placing a personal ad can rent a voice mailbox so that respondents can call in by telephone. Similarly, the advertiser can retrieve the voice message for a fee of one dollar, using a five-digit personal password. Each telephone reply costs one dollar, with Southern Bell retaining 40 percent of the revenue. *May 1990*

Restaurant Reviews

A 900 number -- 900-230-Dine -- planned for 23 US cities will give "confidential research reports" on restaurants, with interactive options ranging from types of cuisine to price ranges.

Callers from hotels, pay phones or some businesses that have 900 numbers blocked, have the option of using an 800-776-Dine and charging the call to Visa or Master Card. The application costs $1.50 for the first minute and 80 cents for each additional minute. *Aug. 1990*

Gourmet Food Via Federal Express

The second annual Dinner a la Federal Express directory was available for the Christmas Season. Containing 160 food companies and restaurants throughout the nation that Fedex everything from barbecue ribs to smoked salmon to your door, was available for a $3 phone call to 900-820-EAT. *Feb. 1991*

New Recipes Just a 900 Call Away

Using a touch-tone menu to guide callers to recipe categories, William Chamberlain, an "I Love-to-Cook guy," gives recipes over the phone.

For 95 cents a minute, a typical call to 900-226-COOK lasts three to four minutes, and the caller may copy the recipes or request them by mail. Up to 16 recipes are available. Callers also have an opportunity to leave their own favorite recipes. *March 1991*

Dinner Refund Offered on Dining Line

The Dining Line in Philadelphia has signed up 40 restaurants and predicts they will have more than 100 by summer. Callers to 900-903-2468 select a restaurant from the various choices given them.

While still on the 900 call, if they make a reservation, they are given a code number which entitles them to one free entree. The cost of the call is between $4 and $6. *April 1992*

900 Creating Psychic Need
Demand Grows for Metaphysical Types

An article in the *Los Angeles Times* described a new growth industry for job seekers. The *Times* described the category as, "metaphysics: an assembly of soothsayers, seers and purveyors of knowledge from the nether realm of thought." They were referring to psychics, tarot readers, astrologers and numerologists. It seems the writer had read a want ad that the Psychic Market Group had placed looking for the above. In particular, it sought those in private practice who wanted to shift to a situation where they would have fewer hassles and no billing.

According to the article, good psychics are in demand and can make up to $18 an hour working for a 900 number. *Feb. 1992*

Travelers Love 900
Meets Info Need

900 telephone numbers are becoming popular with travelers. Even James Herold, Director of IAS Marketing for Pacific Bell admits to calling USA Weather before leaving on a trip -- and from his home phone. He said, "That's an example of quality 900 programming providing a real value."

Using 900, reservations for national campgrounds can be made at 12 parks, including Yosemite, Cape Hatteras, and Assateague Island. Call-cost charges are 75 cents for the first minute and 50 cents for each additional minute.

In Chicago, FastBreak Vacations provides a 900 telephone number that gives information about discounted travel possibilities. Updated as often as twice a day, the recorded information deals primarily with package tours, cruises, hotels, condominiums and air fares. Call-costs are $1 a minute.

Chicago also has a Cruise Shoppers Hotline offering information on discounted cruise packages and specials. Updated weekly, the call-cost is $2 a minute and callers may select from several cruise destinations.

In Florida, Kenney Communications gives such information as motel and recreational vehicle park locations. Call-cost is 75 cents for the first minute and 50 cents for each additional minute. *Oct. 1990*

FlightCall Begins 900 # for Airline Info

Travelers can find out what time their plane is arriving along with other updated flight information by dialing the Official Airline Guide's (OAG) new number, 900-786-8686.

Information usually seen on an airport terminal monitor, such as arrivals, departures, gate and baggage claim area information, is available.

Perception Technology is supplying the equipment. The recording is updated every 10 minutes, 24-hours a day for a call-cost of 75 cents.

The FlightCall system is currently being used in Chicago's O'Hare airport. OAG plans to introduce FlightCall at 14 other airports around the nation by the end of 1991. *Nov. 1990 (active)*

Vacation Rental Properties

Club Bed, of Lake Tahoe, Nev., has introduced a 900 service which allows callers to contact owners of vacation rental properties nationwide.

Customers can also list their own property or propose an exchange of their home or time share on specific dates. 900-CLUB-BED offers the information listing for all states at a cost of 95 cents a minute. *Feb. 1991*

National Spring Hotline

During the last Spring Break, students were able to get the latest beach, party and concert information by calling the National Spring Hotline (900-884-4343). The $2.00 service offered a menu covering the nation's most popular areas.

Beach coverage and reports included Hawaii, California, Padre Island, Corpus Christi, Galveston, Texas and Florida. Updated twice weekly, the service features weather, tournaments and celebrity appearances. *July 1991*

Credit Okayed
Airlines Save Bucks

Travel Weekly reported that when airlines balked about paying 800 bills for travel agents calling for credit card approval codes on cruises they were booking, a 900 number solved the problem neatly. The goal was to try to stop the agents who needed cruise sale approvals from using the toll-free 800 telephone numbers that handle credit card approvals for air sales.

The service is estimated to save the airlines approximately $4,000 a month. Agents who call on the 900 number have top priority since they are "billable by the minute." Kerry Stewart, MasterCard's director of voice services, said. "The average call lasts a hair under one minute, which would cost 60 cents, but this obviously depends on how fast the travel agent reads off the required number." *July 1991*

The Best Vacations On 900

An October '90 *AN* article, "Travelers Love 900," reported on the value of 900 for travelers. A recent addition is Hospitality Ventures, which conducts a biweekly survey of travel providers such as travel wholesalers, large travel agencies and tourist boards. Ventures makes a selection of the three to five best value vacation trips and provides detailed descriptions of them to people who call 900-288-9600.

Designed to serve N.J., Penn. and Del., all the described trips leave from Philadelphia.

Updated every other week, calls to the line cost $1.95 for the first minute, and $1.25 for each additional minute. The program is being marketed through newspaper and radio advertising. Radio does not appear to be working thus far, and although print is doing better, the early results are disappointing.

A thirty-year travel professional, Bob Dixson, president of Hospitality Ventures, designed the application. He said, "My experience allows me to determine which trips are the cream of the crop, the ones that offer maximum value and enjoyment, and I simply tell callers about them on the 900 number." *Nov. 1991*

800 to 900
Tourism Information in Minneapolis

The last few years brought the Minneapolis area a tremendous increase in tourist requests for information -- up 100 percent from 1989-90 and 1990-91. That expansion caused the Greater Minneapolis Convention and Visitors Association to turn to 900 to offset the cost of fulfilling the demand for visitor information.

Until a year ago, the association had used an 800 number to handle requests for travel brochures. In January 1992, they switched to 900-860-0092 at $1.99 for the first minute and 99 cents for each additional minute. AT&T manages the line, and the association said they were good to work with.

Callers hear information on events in the Minneapolis area as well as on the products and services of the nearly 700 members of the Greater Minneapolis Convention and Visitors Association. They can be switched to a live operator to get lodging information while still on 900.

The association's brochures promote the line and calls to its local number are referred to the 900 exchange. "From a service or public relations standpoint, it's not a perfect solution," said Nancy Preston, vice president of marketing. "We would prefer not to charge people for information; we want to attract visitors. But [the 900 line] does allow us to service more visitors per year." The line permits the association to handle a higher volume of calls, Preston said, plus it brings revenue to help cover costs of production and mailing of brochures.

Average calls to the line last three minutes, and the volume of actual call counts varies throughout the year, Preston said. Counts are higher during the summer travel season and whenever Minneapolis hosts conventions. *Jan. 1993 (active)*

900 Number for Last Minute Travel Plans

Last Minute Travel Connection at 900-446-8292 offers information on discounted travel offers for $1 per minute. The line, operated by La Onda Ltd., brings together travelers and travel companies with leftover reservations. *Sept. 1992*

A 3-year Personal Touch
100,000 Callers in '92 to France On Call

France on Call, the French Government Tourist Office's 900 line, in its third year of operation, is doing well and facing the future with plans to deepen the personal service its callers receive.

Callers to 900-990-0040 speak to a live operator for general tourism information on France and the French West Indies for 50 cents per minute. Operators ask specific questions about the caller's planned trip, and that information can be used to prepare a personalized computer-generated itinerary. "In 1992," Michel Bouquier, director general of the French Government Tourist Office, said, "40,000 itineraries were sent out." The itineraries give callers maps, mileage and travel time between destinations and descriptions of highlights.

The service bureau for the line is Next Step Marketing. France on Call originally went on line under another number, but changed that number two years ago when it switched to Next Step Marketing from previous service bureau Europe Assistance.

France on Call generated 100,000 calls in 1992. When it began in 1990, the line received more than 60,000 calls in its first few months (AN, July 1990). An average of twelve operators staffs the line from 9 a.m. to 5 p.m., Monday through Friday. Additional weekend and evening hours of live operation are added during major marketing campaigns.

According to Bouquier, the line plans to begin sending callers personalized follow-up letters giving more details about what there is to see and do during their trips. The secret of the line's success is the quality and personalized service, Bouquier said. "We're not just sending out brochures."

The line is marketed through a lot of networking, Bouquier said. Travel agents, airlines and French restaurants distribute brochures. The line is also mentioned in the tourist office's annual deluxe brochure, France Discovery Guide, and callers to the New York City office are directed to the line for information on travel to France and the French West Indies. *June 1993 (active)*

Biblical Passages Via Satellite and 900

By the end of summer, a religious cable station available via satellite will deliver biblical passages to callers who dial a 900 number. The station, to be called "Worship," will be operated by Christian Network Inc. Developed by the co-founder of Home Shopping Network, Lowell W. Paxson, the station hopes to sign up cable operators, TV and radio stations. *July 1992*

Loving Fans Call Celia Cruz

Celia Cruz, the Cuban salsa queen, possibly the most revered living figure in Cuban culture, has a psychic line. 900-454-CELIA runs $3.99 a minute and the line is the big topic of conversation in Cuban circles. In an infomercial Celia urges viewers to call, and -- they do. *April 1994 (active)*

Fishing Info Via 900

Fishing enthusiasts can call the Oswego County FISH-N-FUN line at 900-933-FISH for fishing reports on that area of New York. A copy of "Trophy! Oswego County Trophy Fishing Guide" is also available on the 95-cents-per-minute line. *June 1993 (active)*

Chinese Astrology

The Chinese Astrology Hotline at 900-443-0308 offers callers access to ancient Chinese astrology based on their year of birth. And it is beginning to make a believer out of Tom Fiorillo, the line's sponsor. Fiorillo, born in the Year of the Snake, has operated the line for about six months.

For $2 per minute, callers can hear predictions concerning romance and career in calls averaging three minutes in length. In this type of astrology, the year of birth gives keys to one's personality, mental attributes, and degree of happiness and success throughout life. Fiorillo was reluctant to identify the service bureau supplying him with the database for the line.

According to Fiorillo, the line is going well. With a monthly advertising budget of $2500, he takes out classified listings and one-inch by one-inch display ads in newspapers across the country. *Nov. 1992*

Dial-A-Chef

Callers in need of culinary help can turn to Dial-A-Chef, a 900 service that provides food preparation advice or a recipe from a library of 100,000 dishes.

The line, at 900-933-2433, is operated by the Ocean City, Maryland-based Culinary Consulting Services. The $2.95-per-minute line offers callers personalized answers to food preparation questions. Pat Nash of Culinary Consulting Services said that if research is needed, callers are asked to call back on the 900 number at a later time.

Nash said that the idea for the line grew out of his work in food service where he repeatedly heard customers asking questions about wine and food. Nash and his wife Karen both still work in food service while operating the line. They handle calls with the help of eight consultants.

The line, which began in November 1992, operates from 9 a.m. to 5 p.m. Monday through Friday Eastern Standard Time, and noon to 4 p.m. on weekends. Washington, D.C.-based Telecompute Corporation is the line's service bureau.

A sister line to Dial-A-Chef is the Wine Line at 900-933-9463 designed to match the ideal wine with a caller's menu and budget.

According to Nash, call counts are still somewhat slow, a fact he attributes to trying to determine the best advertising method and venue. At the time *AN* spoke to him, advertising in TV Guide's regional issues had just started. With the help of its marketing firm, National Television in West Palm Beach, Florida, Culinary Consultants is also using direct mail.

Other publicity includes an article in The Squire, a newspaper in the area of Kansas that Nash is originally from.

Although the lines are operated year 'round, Nash sees them as somewhat seasonal applications, with the winter and spring months bound to attract more calls due to holiday entertaining *Jan. 1994 (active)*

New Horoscope Line

Harper's Bazaar magazine has launched a 900 horoscope line featuring famous astrologer Nicholas Campion. The line, at 900-420-1993, will supplement Campion's monthly magazine column. For $1.50 per minute, callers can hear him read the horoscopes. *Jan. 1993 (active)*

International Travel Info
Health Information For Travelers Available

Travel Health International now offers up-to-date information on health and other topics. For 95 cents per minute, travelers can dial 900-454-7525. The service covers more than 200 countries.

Updated weekly, information includes vaccination requirements and recommendations; overviews of health concerns, including disease risk and suggested corrective action.

Additional topics cover security, visa and other entry requirements, climate, language, currency and geography. The average call lasts approximately eight minutes.

"The cost is a bargain when weighed against the time and effort it would take an individual to gather the same information," said John Dayton, president of the Atlanta-based Travel Health International. According to Dayton, his company monitors sources such as the Centers for Disease Control, the World Health Organization and the U.S. State Department.

Callers also have the option of using a toll-free 800 number and ordering reports by mail at a cost of $7.95 for the first country and $3.95 for each additional country. *July 1992*

Maryland Info on 900

Maryland 900-VISITOR, providing information on events and attractions throughout Maryland, is among the latest travel-related 900 lines to enjoy success.

The $2 per-minute line which began April 1, grew out of an idea that there should be a quick way for people to get information, said

Roseanne Souza, president of the information provider, Souza Productions and Entertainment Inc. Travel kits are available from the Maryland Office of Tourism and Development, but mailing time delay is a problem.

The line has gathered support from Maryland's Office of Tourism Development, Souza said. She uses that state office as one source of information for the line and as a source of promotion. The office's ten information centers all have counter-top displays promoting the 900 number. Souza also advertises in the Maryland Tour and Travel Guide and in the state's Calendar of Events.

Other promotional efforts have so far mainly been public relations. Souza added that billboards on highways leading into Philadelphia and elsewhere in neighboring Pennsylvania are planned. The line has six categories: shows, festivals and fairs, maritime and state parks, professional and college sports, historic attractions and holiday and seasonal happenings, with room allowed for expansion into other categories. The system does not allow calls to run longer than seven minutes.

Sprint serves as both carrier and service bureau. Souza looks forward to expanding the line into coverage of other states, such as Virginia and the District of Columbia, shortening the name to just 900-VISITOR. *June 1993 (active)*

New Singles Travel Line

This month, Lifestyles For Singles™ launches a new singles travel 900 line. Doyal Bryant, the president, said, "I am excited about this service, replacing the Travel Info Line. This application not only helps singles select travel destinations as the old line did, but offers new features including travel personals. Callers can connect with travel companions and travel groups!" *May 1993*

NUDE Number

Anthony and Leanne Cadiz of Boca Raton started a 900 number to access locations, directions and descriptions of beaches, rivers, lakes and hot springs where nudity is allowed. For $1.99 a minute, callers to 900-Sun-Spot can get listings of 379 such places in the U.S., including 27 in Florida.

The 900 number is not about sex, Cadiz says. In fact, callers are encouraged to follow what Cadiz calls "nude beach etiquette." For example? "Ask permission before taking photographs." *Aug. 1993*

New Numerology Line

Jeraldine Saunders, creator of "The Love Boat" television series and well-known numerologist, has launched a numerology line at 900-370-0671. Saunders' Numerology Forecast is designed to give callers consultations on personal matters, such as business and romance. *Nov. 1992*

Recipes Via 900

Another health-related 900 line is the Happy Healthy Gourmet Healthy Diet Line, sponsored by the Happy Healthy Gourmet and Family Fitness Gym in Lancaster, California.

Callers to 900-329-9431 obtain recipes with nutritional information, including calories, fat, cholesterol and vitamins for each. Calls cost $1.99 per minute.

The line also features fitness tips provided by Natural Body Building champion Rory Mosely. *July 1993*

Positive Press for 900
Religious Lines Cited as Worthwhile Calling

The majority of consumers have a negative perception of 900 applications. However, some 900 numbers received a plug in the weekly newspaper *Evangelist*. A good-sized boxed item under the headline "900 Numbers Worth Calling" listed a few applications concerning religion.

Among those: A Catholic "Helpline" at 900-740-7733 gives a message on various topics of faith for 83 cents per minute. A 10-minute homily taped from Sunday Mass at St. Ignatius Loyola in New York runs on 900-740-4664 for $5 per call. A line from the St. Anthony Guild, 900-ANTHONY, offers a two-minute spiritual message at $2 per call. A movie and video review line from Catholic News Service, 900-PREVIEW, costs $1.50 for the first minute and 75 cents for each additional minute.

Also listed was 900-820-POPE with daily two-minute excerpts from talks by Pope John Paul II at $1.95 per minute. The line obtained a new number earlier this year. A separate number, 900-786-5252, offers the message in Spanish.

A new application didn't make the list, but is a good candidate for it. Callers to 900-860-BIBLE can hear Bible readings at $1.95 per minute. *Sept. 1992*

U.S.-Holy Land Prayer Line Begins

A recent addition to religious 900 lines is the direct U.S.-Holy Land prayer line launched by Bezek, Israel's state-run phone company and The Jerusalem Christian Review, a newspaper.
Callers to 900-4-HOLYLAND speak to live operators who pray with the caller for whatever topic the caller names. Operators are also trained to offer crisis counseling. The line began last June. Calls cost $1.99 a minute. *Sept. 1993*

Tammy Faye 900 Line

An inspirational line with a well-known name attached is Tammy Faye Bakker's "You Can Make It" Line at 900-288-TAMMY. The line features messages written and read by Bakker and designed to help callers with life's day-to-day problems.

Calls cost $2.99 per minute, and the messages are updated every Monday. Jonathan L. Weisz, president of the line's service bureau, La Jolla, California-based New Media Telecommunications, Inc., said that Bakker is currently promoting the number on the talk show circuit. According to Weisz, that promotion is pulling calls in, with each mention generating 1,000 minutes.

A friend of Bakker's recommended her to the four-year-old New Media Telecommunications. Weisz spoke glowingly of Bakker, calling her "the sweetest and most genuine person I've ever met." *March 1994*

"Psychic Friends" Travels

Mike Warren is the president of Inphomation, the company that produced the Psychic Friends Network infomercial starring Dionne

Warwick. That infomercial has generated more than three million 900 calls since 1992.

Now the Baltimore-based Inphomation brings the infomercial franchise to the rest of the English-speaking world. Starting with the U.K., Canada, Australia and New Zealand, deals are in various stages of legal review, negotiations, testing and rollout.

In Canada, working with the Florida-based ICN, it will use existing infomercials in the bordering regions with Canadian and US 900 numbers. Additionally it is testing, using the same infomercials, in some major Canadian markets, with Toronto as the first test site.

In Germany, the thinking is to develop totally customized infomercials, although some dubbed tests will run. Future plans include Spain, Brazil and Mexico. *October 1994*

TPI Launches Tele-Personals Network

Tele-Publishing, Inc. (TPI) one of the nation's major providers of voice personals to newspapers, has launched the Tele-Personals Network (TPN), the first totally turnkey interactive cable television network. It debuted June 20 in the Chicago market with Jones Intercable, reaching 200,000 subscribers.

TPN combines interactive telephone and television technologies, allowing ads to be sent electronically to local cable head-ends for presentation on L.O. and/or photo-classified channels. This technology eliminates costly capital investment, staffing and production expenses for cable operators.

Advertisers place their ads by calling a toll-free number, and can also record a voice greeting and retrieve messages for free. Viewers respond by calling a 900 number, just like the current voice personals services mainstream in about 80 percent of the nations major daily newspapers. Revenue is shared by TPN and the cable operator.

Jones Intercable, the first cable operator to carry TPN, is the country's seventh largest MSO (multiple system operator). Launching just about every week with other cable systems, TPN was reaching approximately one million cable subscribers as of last month, said Cindy Reid Dorsey, director of sales and marketing.

Dorsey added that TPN expects 1.5 million subscribers by the end of September. She reported the network going strong, performing best in the big city markets, like Chicago and Los Angeles, with their large concentrations of singles.

TPN recommends that cable networks run the personals programming four hours per day, seven days a week, preferably between 10 p.m. and 2 a.m. The programming has grown to 500 personal ads carried. There is consistency built in with men's and women's ads running in set time segments.

Regarding call counts, TPN's initial projection was 10 call minutes for every 1,000 subscribers in cable system per week, Dorsey said. "For example, with a system that has 200,000 subs, you'd be doing 2,000 minutes a week." According to Dorsey, their projections were met in just the first few weeks of operation.

Marketing toward advertisers includes TV spots that TPN provides to cable operators. Because TPN's personals programming is basically driven by women's ads, Dorsey said, the TV spots are geared toward women. "Get the women's ads and the men will call."
Sept. 1994

900 Personals on the Internet
Electronic Publishing & 900 Numbers

Internet Personals, Inc. has launched a merger of electronic publishing and 900 numbers. On June 17, the New York-based company began making personals databases of publications accessible from the Internet, the vast computer network.

The service, with the Internet address of personalsusa.com, went online listing personals carried by New York-based TRX, with several more bureaus due to add their publications' personals shortly, said Richard Firestone, Internet Personals, Inc.'s systems manager. The company is dealing only with service bureaus, not directly with publications. Internet users who access the personals are not charged, and revenue is generated from calls to the 900 lines.

"For each call that comes in on the new Internet 900 numbers, we share the net proceeds 75-25," Firestone said. The 75 percent goes to the service bureau and its client publications, and the 25 percent to Internet Personals.

The advantages for service bureaus and publishers include increased exposure for their personals ads, Firestone said. He noted that the Internet has approximately 20 million users, falling mostly into the 18 to 45 age group. The combined singles and social and sex discussion special interest groups on the Internet log more than 3,000,000 calls per month. Also, an additional five million computer users have access to Internet e-mail through an on-line service, such as CompuServe.

To reach computer users, the service is being widely publicized on the Internet itself on appropriate discussion groups. Once a computer user gets into the service, he selects a geographical area and then chooses from another menu of publications. Once a publication is chosen, the entire personals section is downloaded into his/her computer. Users who choose to subscribe to the service can get their newly issued personals automatically delivered to their electronic mailboxes.

Firestone is enthusiastic about the service, and thinks that there is great untapped potential in linking 900 applications and the Internet, due to the large number of individuals the Internet reaches. His company is interested in talking to service bureaus with ideas for other marriages of the Internet and 900.

But his company's personals service does face one problem. "There is just so much text that we can carry on-line at any given time," Firestone said. He explained too much text on a server increases download times, defeating the entire purpose of providing users with quick access. *July 1994*

What's Going on in Branson?

According to AT&T, the Branson Lakes Area Chamber of Commerce in Branson, Missouri has one of the most successful 900 travel lines in the country. People from all over flock to Branson to hear Mel Tillis, Bobby Vinton, Glen Campbell, Andy Williams and Boxcar Willie, among others, perform in their own theaters.

Callers to 900-884-2726, for $1.50 per minute, get an overview of what's happening in the Branson Lakes Area, and are given the option of leaving their address or fax number to receive show schedules, vacation packages, lodging information and area maps showing alternate routes that avoid the traffic congestion Branson is becoming famous for.

During the season, from May to October, the 900 line gets up to 100 calls a day over 8 incoming lines dedicated to the 900 program. During the off-season they will get 10 to 20 calls a day. *Feb. 1995*

Travel Promotion Via 900

The Kitsap Visitor and Convention Bureau in Washington State now uses a 900 line, Prime VIP, at 900-773-3836, to promote attractions and events in the Puget Sound region. For $2 per minute, callers can choose from a menu of messages on events, areas, activities and attractions. The line is marketed in Washington, Oregon and Northern California.

The Seattle-King County Convention and Visitors Bureau also use Prime VIP lines to promote its areas. *August 1994*

900 Wedding Consultation

Couples in search of a different sort of nuptials can call Weddings in the Rockies at 900-329-1144. The company is a consulting and coordinating service for out-of-town couples who wish to travel to Banff in Alberta, Canada for their weddings. The 900 line offers consulting for $1.49 per minute. *August 1994*

In-Depth Profile
The Personal Dialogue Dating Line

This is one of the very first dating programs to be launched in the country, established in 1988 by TRX Corp. (212-644-0370), headed by Gary Glicker, president, based in New York City. The program can be accessed by numerous 900, 540 and 976 pay-per-call numbers throughout the country. One of the numbers is 1-900-988-5030 ($1.70 per minute). The call charge varies between $1.50 and $1.95 per minute, depending upon the geographic area served and the version of the program.

This program was actually launched before the days of pay-per-call, on regular phone lines, when TRX began handling personal classified advertising via telephone (now commonly known generically as "voice personals") for *The Village Voice*. This program used a basic voice mailbox system for placing and responding to voice personal ads, and it was an instant success. It became immediately apparent to Glicker that this application had some real potential in the emerging pay-per-call industry, which was just getting started in New York with the introduction by New York Telephone of 540 dialing in March of 1988. 900 numbers were yet to be introduced by the long distance carriers.

The Village Voice soon changed over to pay-per-call, and TRX handled its personals program until 1990 when *The Voice* purchased its own call-handling equipment. TRX and Gary Glicker in fact helped get started some of the major players in the voice personals arena, which is now one of the most profitable 900-number applications around.

The program itself functions like most other voice personals programs. The main menu gives four options:

1. Browse the ads by age and gender (the basic program is divided into male and female by age group: younger than 20, 20-29, 30-39, 40-49, 50-59, and over 60).

2. Respond to a specific ad by extension number (printed along with the ad in the newspaper and given verbally along with the voice ad when browsing the program)

3. Pick up messages or record a message

4. Instructions on using the system

According to Glicker, in the early days, when these programs were still quite novel, it was possible to charge callers at every step of the contact process: when placing an ad, retrieving messages, and responding to an ad. Although all of these charge options are still available, in any combination, the clear trend seems to be toward charging for the response only. The placing of ads and then retrieving responses is free in many of today's voice personals. Virtually all of TRX's programs are set up this way. "Without advertisers there are no respondents," says Glicker, and call volume is driven primarily by the respondents. It became apparent that it pays to encourage as much advertising as possible, and giving it away is obviously very good encouragement.

TRX really doesn't do any marketing for callers itself. Rather, it teams up with newspapers in a revenue-sharing partnership. TRX actually offers a complete "turn-key" service for its newspaper clients. TRX maintains the equipment, performs necessary programming, collects all the ads, contacts the advertisers, monitors the program, provides customer service, edits the copy, typesets the pages, and delivers camera-ready pages to the newspaper, numbering anywhere from one page to 8 pages of classified ads. The newspaper simply includes this section as a regular feature and collects a share of the revenue generated by the respondent's calls.

When TRX was just getting started in the pay-per-call voice personals business, it was very hard to convince prospective newspapers to print the Personal Dialogue section for a share of the revenue. After all, these were the days when such personals were the sole purview of a few alternative lifestyle newsweeklies, and the more mainstream papers needed some convincing to take the gamble. TRX was forced to take out ads in these publications in order to demonstrate how popular (and profitable) such a service could be. Eventually these newspapers started seeing the potential and began entering joint venture relationships with TRX. The newspaper contributes the space in its publication and TRX contributes the program and all services necessary to administer the program.

TRX today has offices and equipment in Chicago, Boston and New York, which are among its major markets. The reason for multiple locations is because it's less expensive, at least for high volume applications, to take advantage of the local or regional pay-per-call services from the local carrier whenever possible, but this can be accomplished only with call-handling equipment in-place in that locality.

TRX occupies a large suite of offices in Manhattan, with all voice processing equipment on the premises, consisting of two voice response units (VRUs) of 64 ports each, and with essentially unlimited expansion capability. Its satellite offices in Boston and Chicago each have one 64 port VRU. The hardware and software were developed by TRX and are proprietary.

TRX now handles the personals classified programs for more than 200 newspapers, which run the gamut from small weeklies to special interest newspapers to large daily newspapers throughout the country. The number of newspaper clients continues to grow steadily.

Although reluctant to reveal hard call volume numbers, Glicker indicates that TRX has received in excess of one million calls since the program was established. Not surprisingly, most of the call volume comes in between 7 and 11p.m., when respondents are at home. Some of the newspapers using the TRX Personal Dialogue program are *The Pioneer Press* in Chicago, *The Stamford Advocate*, *The Greenwich Times*, and *The New York Observer*, to name only a few.

TRX actually runs more than one program. Besides the obvious heterosexual categories, by age and gender, TRX runs some specialized programs categorized by sexual preference (gay), race, language (Russian), nationality (Irish), and religion (Jewish). These programs are for special interest publications that are targeted to specific niche audiences. TRX is able to custom design a program for any lifestyle or demographic category necessary in order to respond to the specific readership of the host newspaper.

According to Glicker, TRX runs a "tighter ship" than many other personals services. First, it usually asks for a written version of the classified ad from the prospective advertiser, usually by providing a fill-in-the-blank coupon in the newspaper for the advertiser to fill out and send in. Then a customer service representative at TRX will contact the advertiser by letter to fully explain the system and what will be expected of him or her. It is important for TRX to ascertain whether or not the advertiser is sincere, which is essential with a free advertising program in order to weed out the cranks and to maintain a quality service with legitimate advertisers.

"We take great pains to ensure that everyone advertising is sincere about it, and then we also insist that everyone record a message on the system in their own voice," says Glicker. After receiving the written ad, TRX immediately sends out a letter explaining the program, how to record the ad in the advertiser's own voice, and how to access responses to their ad. If necessary, in cases where the advertiser has not recorded his or her ad or otherwise has not followed the letter's instructions, TRX will often follow up with a telephone call to make sure the letter was received and to encourage participation in the program.

One of the reasons behind the long-term success of this program, according to Glicker, is the extra effort in customer service and follow through. For example, if they notice that an advertiser has not collected her messages for awhile, someone will actually call and remind her. "There is no question that we police our program more carefully than the others," says Glicker. As a result, they keep customers for a long time and get a lot of repeat business from satisfied advertisers and respondents.

A unique feature of the Personal Dialogue program is the non-use of any abbreviations in the printed ad copy. No need to decipher what SHCNSM means (single, Hispanic, catholic, non-smoking, male?). Just introduced is a new feature that allows direct contact between the advertiser and the respondent, named "Call Direct." If the advertiser elects to have this feature, the call will ring through to his or her designated phone, during a designated time period only, and the advertiser is given the option (interactively by keypad input) whether or not to accept the call at that time. If accepted, the two parties are connected. If not (too busy right then, a mouth full of pasta), the caller will get a message with words to the effect, "Your party is unavailable at this time, but please leave a message and he/she will get back to you as soon as possible." This option will be attractive to women who are reluctant to give out their telephone numbers to people they don't know very well yet, because they can always be reached through the program without having to ever divulge their actual phone number.

An important point for newspaper publishers to keep in mind about pay-per-call voice personals, according to Glicker, is that the paying customer is no longer the advertiser. The respondent to the ad is actually the customer. With the traditional print classified advertising, the newspaper "never felt any responsibility toward the respondent, it was always toward the advertiser, and all their efforts were geared toward the advertiser because the advertiser was the one paying the freight, and he or she was the repeat customer," says Glicker. Although voice personals look the same in print, the true customer is the respondent, and it is he or she that will become the repeat customer and who will generate the revenue for the program. "Unless you focus on the respondent," continues Glicker, "chances are over any period of time these columns will come and go or fail as quickly as they start, because too many people focus on the advertiser. Give the respondents some value for their money, and they'll support you."

Newspapers now see these programs as more than another revenue generator, but also as a service to their readers. And if the competition has such a section, it had better get one too or risk losing readership. Some savvy publishers will use this section as an anchor for display advertising sales that relate to the singles market: bars, singles cruises, restaurants, clubs, concerts, clothing stores, and anything else that appeals to singles.

For fairly obvious reasons, the vast majority of respondents to voice personals are men. In our culture men have always been expected to take the initiative in pursuing women, and it's no different in the voice personals. According to Glicker, between 60% and 75% of the callers to his voice personals system are men, depending upon the geographic area and the host newspaper readership demographics. The goal, therefore, in any successful program will be to attract as many female advertisers as possible in order to continue generating calls from men.

The average call duration for respondents to Personal Dialogue is about 5 minutes, which is considered quite good by industry standards for this type of program. This seems to indicate that the callers are quite satisfied with the program itself, and that there are sufficient legitimate and appealing ads to respond to, a rather critically important feature for any such program.

Having been in the business now for several years, TRX now has an extensive database of voice personals advertisers and former advertisers in numerous market areas around the country. To encourage advertising by female advertisers in particular, TRX will on occasion conduct a direct mail promotion to encourage these women to advertise again in their local host newspaper.

Has the Personal Dialogue actually worked to bring people together? The former publisher of a major New York weekly newspaper met his current wife through the Personal Dialogue section in his own newspaper.

Chapter 10
Education, Careers & Self-Improvement

In the early days of 900, when there was little regulation in place, some of the job search lines were the worst examples of sleaze, preying upon unfortunate people who were desperate to find work. Along with the phony sweepstakes and easy credit lines, these lines helped to give the 900-number industry a slimy reputation that it's still trying to shake off.

Fortunately, through federal legislation and self-regulation, the industry has been cleaned up considerably, and the majority of the programs offered today are perfectly legitimate.

TCU Provides Advice
Radio Station Promotes 900 Number

Telephonic Connections Unlimited (TCU) provides a 900 informational and inspirational hotline targeted to the African-American community, that is promoted through the WWRL radio station's Better Life Team.

Callers to 900-454-1111 receive advice, tips and "how tos" on getting promoted on the job, wills, credit cards, retirement, budgeting and other money related topics, as well as inspirational messages.

The messages are provided by well-known gospel artists, famed evangelist Dorothy Norwood and health specialist Dr. Gerald Deas.

The newest member of the team is M. Gasby Greely, a TV personality and business advisor. *Feb. 1991*

Colleges Use 900

Universities across the country are using a 900 number, 900-420-1212, to inform future students about Campus Life.

Callers hear a three minute message about any of the universities' offerings, such as academic programs or other data usually received in campus visits. Calls cost $2.50 and each school has its own four digit access code. *Nov. 1990*

School Closings by 900 Number

Parents and students of Northwest Suburban High School District, Illinois, had a new service this winter designed to improve communications regarding school closings. For a 95 cent phone call, they could get that information immediately without having to wait for radio announcements. *April 1991*

Gossip Line

Don Fitzpatrick, a San Francisco based news headhunter, launched a new telephone service. It's an outgrowth of his *Rumorville USA*, a computer newsletter, which is a gossip sheet on the broadcast industry.

The application, called Pipeline, 900-456-2626, is updated daily by news directors across the country and also provides coaching on such basics as doing standups and dressing for television success. The call costs $1.95 for the first minute and 95 cents for each additional minute. Voice mail allows the caller to skip around job-listings, job-hunting and Rumorville tapes. *May 1990*

ACI, FUJI & EDWARDS

Included in each box of FUJI 5 1/4 inch computer disks is a Home Office Guide which includes the Hotline number, 900-456-WORK.

Audio Communications Inc. (ACI) in partnership with home office experts, Paul and Sarah Edwards, brings "how tos" to workers who are making the transition from the office to working at home.

The cost for the call is $2.00 for the first minute and $1.00 for each additional minute. *June 1990*

USA Using 900 Number
Employment Line Helps Get Workers

While the FTC, FCC and Congress are investigating 900, the U.S. Office of Personnel Management is using 900 to promote America's biggest employer--the United States Government.

Callers to 900-990-9200 obtain current employment and career information, application forms and materials related to the Administrative Career With America Program. The interactive line costs 40 cents a minute and has a menu of topics relating either to the college hotline or the general job line. For those of you who are wondering, yes, there is a preamble with a "kill message." *Feb. 1991*

Overview Of Job Markets
New Job Line Uses DOL Forecasts & Studies

An unfortunate experience with a service bureau didn't keep Dr. William E. Meyer out of the 900 business. His company, TriComp, is on the scene with an employment line, 900-420-Jobs.

This is not the usual "looking for work" line but a forecast of the employment scene in terms of demand for the balance of the 90s. The forecasts are based on studies, analyses, and projections by the Bureau of Labor Statistics. The forecasts take into account how the economy is reshaping, foreign competition, and shifts in demands for goods and services.

The line is designed for people looking for new careers and contemplating career changes. For $1.95 per minute, with the average call projected to be 6-9 minutes, callers can hear about new job areas.

Dr. Meyer disclosed that TriComp will promote the line on radio and in print, and use a 30 second commercial on the networks and cable. *Oct. 1991*

Employment Line Hangs On

Although employment lines have taken a beating in today's regulatory atmosphere, financial industry job seekers from across the nation still can find job leads on 900-773-7678.

The IP is a former investment banker, Gary Salton, who is working with Andy Sutcliffe of Tele-Publishing, Inc., to deliver the service. For $1.95 for the first minute and 95 cents an additional minute, callers have access to more than 1,000 openings.

Salton gets the job openings through 15 national associations and institutions including The Association of Management Accountants and the National Association of Certified Fraud Examiners. The associations help promote the number through their mailings, newsletters, and ad space in journals, in exchange for which Salton provides a share of the profits.

Additionally he maintains standing relations with 250 search firms who have job listings. Salton said, "Despite the fact that since April there has been only one complaint and that was handled, AT&T keeps us going through hoops for authorizations and certifications."

When *AN* questioned Salton about the success of the venture, he said, "not great." *Feb. 1992*

Talent Search Uses 900

Out of Ogden, Utah appears a 900 HotNews line for the gifted and talented. Individuals are requested not to call if they are not the best at what they do. The purpose -- to bring together individuals who have extraordinary expertise. *Nov. 1990*

Business & Student Job Line

Dr. Donald Casella, director of the Career Center at San Francisco State University, is confident that the center's job listing service became "ten times better" after a switch to a phone system involving 900.

Businesses call a 900 number and dictate a job listing costing $5 for the first minute and $2 for each additional minute. Students call a local number to listen to the listings.

The 900 line, begun two years ago, receives about 400 calls a month. The average call is two minutes. The university contracted directly with Pacific Bell.

Casella said, the speed and simplicity of dialing in and dictating a job listing pleases businesses, and the cost of the 900 call is considerably less than the average $76 charge to run a two-line classified in the area's major newspaper for a week.

Prior to the 900 line, the Career Center's bulletin board listed the jobs businesses called or mailed in. The use of the 900 number cuts down on the number of staff and reaches more students, Casella said. Anywhere from 800-1,000 students daily call the local number to listen to the job listings.

Casella said the 900 line also adds a personal touch to the job listing service, because students hear listings in potential employers' own voices. *June 1993*

Buy Freedom Network
900 Benefits First Loan Recipient

Maxine Wilcox, a 52-year-old mother of six from Sacramento California, has been named the first recipient of a loan guarantee from the Buy Freedom 900 Network. The network, launched in January (*AN*, Dec. '91) by television commentator and nationally syndicated newspaper columnist Tony Brown, is designed to assist small businesses and the economy of urban areas nationwide.

The network offers informational services, including the Self-Help Directory, 900-976-6670, a listing of community businesses from across the nation, at $1.99 for the first minute, and 99 cents for each additional minute. Profits from calls to the Self-Help Directory are used to make loan guarantees for small businesses, such as Wilcox's bridal dress business, "Wedding Creations."

The national program has a localized focus. The profits earned from calls in a specific city or area will be put toward loans to start or expand businesses in that area.

Wilcox's $8,000 loan from the United Bank of Philadelphia will enable her to move her business into a storefront, and ultimately to hire her two daughters, both of whom have children and are on public assistance. Wilcox herself has been supported by Social Security

Disability payments, and was previously refused a loan by 12 banks in Sacramento and the regional SBA, because of the modest amount she needed and a negative credit report.

Last month, when Brown introduced Wilcox to the general press at a New York City luncheon, she charmed everyone in the room. However, the story was not covered in the papers. In conversation with *AN*, Tony Brown expressed the opinion that the story was not picked up because it is a positive marketing concept for Afro-Americans. *AN* suggested an additional possibility: the story wasn't picked up because it was a positive 900 story.

Callers to the Self-Help Directory are offered discounts on any service or product, a guaranteed refund and prompt order fulfillment, and are under no obligation to buy. In addition to businesses, religious and non-profit groups are also listed in the Directory.

The program is designed to build up the marketing muscle of existing Afro-American businesses and help create thousands of new ones. Brown believes that the color of freedom in America is green. His focus is on economic development among Afro-Americans.

"While the 'Buy Freedom' concept is not based on the philosophy of 'buying Afro-American,' it is based on the belief that poverty is the number one problem that Afro-Americans confront, and therefore they must become, as a community, economically self-sufficient," Brown said.

A national foundation, the Self Employment Enterprise Fund, Inc. (S.E.E.F.), administers funds donated by the network and guarantees loans to the recipients selected by a committee. It is estimated that approximately 50 cents of each average $3 call will be set aside for loan guarantees. Each recipient is required to take a training course offered by a local, state or federal agency, or a non-profit group, and to develop an effective business plan.

The network operates on the AT&T DEFINITY Communications System, an advanced digital premises-based communications system.

In addition to the Self-Help Directory, the Buy Freedom Network consists of separate 900 numbers for the Tony Brown Hotline, The New Republicans Agenda, Buy Freedom 900 Opinion Poll, and Opportunities/Entertainment. *July 1992 (active)*

Dialing Campers

Dialing 900-CAMPERS will connect you with Camp USA, a business owned by Michiana, Indiana resident Sharon Love. The number is a clearinghouse for information about camps and clinics for students wanting to hone athletic or academic skills.

People interested in obtaining information about camps or clinics can make a single telephone call at a cost of $9.95, and receive information by mail. *Aug. 1993*

Musical 900

Musical groups in need of new members can call the New Musician's Network at 900-454-2JAM. For $3.99 a minute, callers can listen to audition recordings of players from various musical genres.

If interested in a certain player, they can obtain a contact phone number. *Nov. 1993*

Musicians Work Line

Business-related 900 applications touch on many different businesses. An unusual one is Auditionline for professional musicians. This year-old application, the brainchild of two musicians, faces the future with solid plans for increased marketing and expansion. IP Anthony Steele, co-owner of Musicians Networking International Inc., the application's parent company, said, "We designed Auditionline to be a one-stop shop for working musicians, both those looking for work and those looking for other musicians to work with." The convenience of a phone call and the ability for constant updating led Steele and fellow musician Tom Polifrone to think of 900.

Steele noted that the traditional print advertisement route of finding work or other musicians is subject to delays due to publication's printing schedules. "We knew there had to be an easier way," he said.

"We feel that what it has to offer is unique." Steele added, "We utilize two 900 numbers." Musicians looking for work call 900-77-2LIST, leave a 45-second ad and pay $4.99 a month. Bands or producers looking for musicians call 900-285-2JAM to listen to ads for $1.99 for the first minute and 99 cents for each additional minute. That line also features a Music Services category with ads from producers, recording studios, managers, publicists and photographers.

With little to no advertising, the call counts are about 1200-1300 per month. The application's service bureau is East Meadow, New York-based Automated Voice and Fax Solutions. Steele cited the bureau's painstaking efforts in getting Auditionline up and running.

Past marketing efforts have included print ads in publications such as *The Village Voice, Guitar World* and *Musician.* The marketing plan for the near future includes more print advertising, joint ventures with publications, celebrity endorsements and an infomercial.

Other plans include making Auditionline an international application by adding European musicians to its callers and database. Also, at the time *AN* spoke to Steele, his company had investment groups interested in the line. *Feb. 1994 (active)*

Writer's Helpline Still Going Strong

Some 900 applications are short lived, but the *Audiotex News* list of lines that operate after more than two years continues to grow. The Writer's Helpline, as a for instance, at 900-988-1838, ext. 549 is now in its third year of operation (*AN*, October '92).

The line, which began in April 1991, offers news and tips useful to writers for $2 per minute. This includes notices of new literary agencies or publisher's imprints, book packagers' requests, calls for writers with particular skills, and writing tips and advice; all gathered by IP Marion R. Vuilleumier, a writer, teacher and consultant. Callers can also mail in questions to be answered on the line.

The line's content is updated weekly, a factor which Vuilleumier said gives it an immediacy not found in writers' magazines or newsletters. To promote the line, Vuilleumier relies on press releases she sends out to writer's publications; at last count she knew of 90 of those. She also includes the line's number on the business cards and brochures she uses for her writing and consulting business.

The writer's Helpline is not Vuilleumier's only 900 venture. She rents three other lines from Strauss Communications, the same service bureau she uses for the Helpline. These are a sports line (900-988-2808, ext. 237), dateline (900-622-0024, ext. 480) , and a live astrology line (900-476-7500 ext. 529).

Vuilleumier also plays the role of 900 advocate. In her column in the November 1991 issue of *Seniors Cape Cod Forum*, she published an article entitled "900 Numbers Are Not All That Bad." In it she noted that the 900 industry grew out of people's need to obtain specific information quickly and easily, and that consumers don't mind paying for good information reasonably priced. She urged readers to keep an open mind about 900. *October 1994 (active)*

Chapter 11
Entertainment

This chapter covers one of the two major divisions of 900-number services: information and entertainment. Just like print or television, the telephone is simply another delivery medium that is suitable for a wide variety of entertainment applications.

To be included in this chapter, the actual purpose or content of the 900-number program is to entertain; or, the line helps the caller in choosing or buying entertainment, such as a music review line. Adult lines, or the so-called dial-a-porn lines, are also included in the entertainment category -- adult entertainment. Because there has been so much confusion and misinformation about adult 900, we have included a separate section on adult programs at the end of this chapter, including an in-depth profile of an adult IP.

The "Perfect Match" Debuts
ESPN 900 Interactive Sports Show

When Stephen Bilkis, 28, woke up on Columbus Day in 1992 in his Long Beach, New York home he had no idea that two years later he would be the producer of what is being seen as the hottest new ESPN property.

What follows is the journey of a dream to a reality and what path Bilkis took to produce the show - *The Perfect Match* - premiering October 3 (1994). Scheduled to air Monday through Friday on ESPN at 6 EST, the program marries game show excitement to 900 technology.

Bilkis, previously a comedy writer on the *Cosby* show, was a writer for NBC sports who had a strong interest in game shows stemming from his appearance as a contestant on *Wheel of Fortune* while a senior at Tulane University.

In 1988 Bilkis came up with a game show concept which he refined with his friend Terry Cashman, a singer/songwriter, who wrote the words and music for "Talking Baseball" and presently has a new Sony album.

Bilkis recalled, "I knew this was a great game show and I wanted it to have a 900 interactive segment. I readily admit to being influenced by Carol Morse Ginsburg, *Audiotex News* editor, who helped arrange a meeting with Arthur Toll. At that time Toll was president of Gateway Communications, a service bureau in Pennsylvania. Toll thought my concept was a sure winner and offered to finance half the pilot and the project if we could get air time."

Early in 1993, Cashman arranged a meeting where he and Bilkis pitched the show to David Zucker, ESPN programming vice president. Bilkis said, "We walked out of that meeting with a commitment from ESPN to present and finance part of the pilot."

In November and December of 1993, Bilkis went to Los Angeles where the pilot was shot. He describes long days and nights where he did all kinds of producer-type stuff like: holding auditions for hosts and a Vanna White look-alike, finding audiences, selecting a director, building sets, crewing, getting studio time, graphic designs, editing, facilities, insurance, staff writers, casting directors, contestants, video wall, interactive game finish, wardrobe deals, and a good caterer!

During "play break" in the *Perfect Match*, home viewers get the opportunity to compete for prizes. Speed and sports knowledge are essential elements of the game. There is a flat fee of $4.98 for the 900 call to play each game, and the home contestants compete with each other to see who finishes first with the most correct answers. Looking at a square board, contestants have a multiple choice portion that matches their touch-tone keypad. Some examples: college nicknames have to be matched with the university they represent; matching the NFL team to its city; matching the sport term to its sport.

ESPN loved the pilot, the audience loved the game, and the contestants played well. But dark clouds loomed on the horizon. Toll, then the CEO of Regal Communications (resigned last month), who had committed to finance the project, ran into a problem *(AN, Feb. '94)* with the SEC, his shareholders and Coopers Lybrand. Its funds to finance the show were no longer available.

"Cashman and I were determined to go forward, so with help from ESPN we raised the money," said Bilkis. "We had another problem. Originally Gateway was the service bureau for the project, now we needed a new one."

Bilkis met with David Van DeVeer in New York and Tom Barker in Omaha and chose West Interactive as the service bureau. West loved the idea of doing an interactive show for ESPN and with the help of Nancy Berger, Kristin LiaBraaten moved the project forward quickly. ESPN staffers, Kathy Yancey and Peter Englehart worked with Bilkis to finalize the program and solidify the deal.

Bilkis concluded, "It is truly hard to describe what it took to get this show off the ground. My advice to anyone thinking of doing this is to go to dental school. But I am pleased that the technology we developed has created a buzz in the industry. Who knows whom we will be talking to next?" *October 1994*

Madonna Too Hot to Stop

Bruce Baron of Berkeley, Calif. has started an unauthorized Madonna line. He fully expects to hear from her but for now runs a disclaimer saying that 900-98-Vogue is not endorsed by Madonna and is just meeting the public's demand for Madonna news.

The hotline has Madonna music, concert and movie news, classified ads, plus callers can leave questions about the star to be answered at a later date.

Ten days after the line had started there were 900 phone calls at $2 for the first minute, $1 for each additional minute. *Oct. 1991*

Rocker Line

Rocker Steve Miller, the 60's radical, has a 900 number. According to Miller, "They can call 900-454-STEVE. And instead of getting a rap song and an advertisement for T-shirts, they can press a button and learn how to save the national forests, what to do about nuclear waste, how to deal with toxics and where to go for voting registration. Then they can leave their name and address, and we'll send more information and a guitar pick or something." *May 1992*

Sweeps For Space Travel
$2.99 Call Gives Chance
for 7-day Mission With Soviets

Space Travels Service Inc. believes they will make millions with their new sweepstakes contest application. They will have to, because they are paying the Soviet government $12 million to carry the sweepstakes winners into space.

Callers to 900-258-2MIR hear a different message every week concerning space and can register their names for the random drawing.

There will be 11 monthly winners trained in the Soviet Union, who will be back-up astronauts and win $400,000 in cash. Another 120,000 entrants--10,000 each month--will win a poster. The grand winner has the option of preflight training in the Soviet Union, which will include physical training and Russian language lessons or $12.5 million. *Feb. 1991*

Fantasy Park Sweepstakes
Fox Pilots What May Become a Weekly Series

What's in the Omaha water? Call Interactive bagged another "biggy" -- Fox Broadcasting. On April 13, Fox Broadcasting unveiled Fantasy Park, a fully interactive television sweepstakes in which viewers will vie for prizes ranging from a job as a music industry executive to a "six-pack" of classic American cars.

Viewers registered as contestants, either by calling 900-436-4Fox or through the mail, and winners received their awards on the show either live or on video-clip. Fox says if the pilot on this single airing draws enough audience, it may become a weekly series.

Before they closed, the lines received 177,000 calls, and even with Call Interactive's 10,000 lines, there were "busies." The special lines ($2 per call; a percentage of proceeds to various charities) were open April 1-7 for registration for the two $10,000 cash gifts, given away live during the April 13 show. *May 1991*

900# Guarantees 5-year Audiences

In 1980, WGN-Channel 9, Chicago froze the ticket waiting list for its popular Bozo show because there was a 10-year backlog of requests for tickets.

In 1990 when the list was reopened via a 900 number, 140,000 free tickets were gone in five hours and 20 minutes. Thousands of callers jammed the 900 line seconds after the phone number was broadcast at 12:05 during WGN's coverage of the downtown St. Patrick's Day Parade. Those who got through could leave their names, addresses and phone numbers via voice mail. Confirmation of orders will be received by mail. *May 1990*

Two Million Calls

It's hard not to be starry-eyed about the New Kids On The Block when they've rung up over 2 million calls since they started the 900 line this February. Even *Forbes* magazine is impressed that $4 million of the group's estimated $30 million gross in the past two years comes from their 900 line. It appears that $2 a minute, for the first minute, and $1 per minute after that can really add up. *July 1990*

CANS Airs Line

The Catholic News Service (CANS) has recently introduced a new telephone service called The Preview Line. By dialing 900-Preview, callers receive reviews and classifications of movies currently showing at theaters and available at video rental stores throughout the country.

The line was developed by Advanced Telecom Services, Inc. (ATS), using AT&T MultiQuest®. Thomas Coffey, president of ATS, emphasized that the advanced technology incorporated in the program makes it easier for callers to access their choice of review. Coffey said, "ATS is the only 900 Service Bureau in the entire country offering VoiceTone, enabling rotary dialers to electronically access interactive menus."

According to Thomas N. Lorsung, director and editor-in-chief of CANS, "There are a lot of movie choices and we want to make it easier for people to make sound judgments about them." *Dec. 1990 - Jan. 1991*

Soaps, 900 & Episodes

As reported in the July '90 *Audiotex News*, Episodes is a bi-monthly devoted to ABC Television Network soap operas and 900 was an outstanding engagement.

By calling a 900 number, TV viewers ordered a subscription for $3. Now the year-old Episodes has a rate base of 1.85 million, a paid circulation of more than 2.1 million and its subscription price raised to $4. Its first anniversary issue carried 26 ad pages.

Now if they do well with renewals, we wonder what magazine will be next to team up with 900? *May 1991*

Listen: Then Buy Ticket

San Franciscans can hear a recorded sound bite of Bay Area musical events ranging from classical to folk and country performances. For 75 cents a minute, callers to The Event Preview Line (900-844-4448) hear two segments of a future performance.

The service is updated weekly and has been receiving about 40 calls a day according to Peter Rudolfi, an area musician, composer and promoter. *Dec. 1991*

High Stake Games
Winning Wall Street

A Canadian company is offering a cash award of $25,000 for would-be investors whose portfolios have the best performance during a three month period.

Winning at Wall Street produced by Duke Enterprises costs $49.95 for the game and related materials, plus $3.50 for a three-minute phone call.

After the purchase of the game package, callers receive a membership number, and from that point on they use the 900 number to conduct all transactions. Employing a hypothetical $250,000, they can trade on margin and purchase future contracts. *Nov. 1990*

NY Times Values 900
35 Down 43 Across By Phone

The New York Times has introduced a dial-a-clue phone service for its crossword-puzzle fans. For 75 cents for the first minute, and 50 cents for each additional minute, you can find out the answers for up to three entries from a current puzzle.

Nancy Neilsen, *The New York Times'* head of corporate relations, reports that although still in a test mode *The Times* is so far been pleased with the volume of calls received by 1-900-884-CLUE. *April 1990 (active)*

The Game
May Change 900 Image

The 900 business is going to get a shot in the arm with an advertising blitz that may change the whole 900 scene. This ambitious new 900 telephone service started on May 7, offering callers a chance to win $1 million if they can be the first to answer trivia questions by making a $2.99 phone call.

The application--called The Game --has big bucks behind it with a first month national ad campaign estimated at five million dollars, and $100 million to be spent before year-end.

According to Lynn Folse, the vice-president for 900 Million Productions, the company behind The Game, day one had surpassed all their expectations. "This is a breakthrough for the 900 industry."

MCI is prepared on a day-by-day basis to upgrade the present capacity of two and one-half million calls per day (1736 calls per minute).

Bringing the application to the market are three advertising agencies: Della Femina McNamee, A. Eicoff & Co. and Lois/GGK. Three separate TV campaigns, consisting of both 30 and 60 second spots, will run during all day-parts on national cable, and in five primary spots and 35 support markets in the first few weeks. A sophisticated tracking system will determine on a daily basis which commercials are most effective.

Home Shopping Network provides the telephone operators and the automated voice response units through its wholly owned subsidiary, Precision Software, Inc. Prospective players dial

1-900-226-2000 and are matched up with nine other players. When *Audiotex News* played the game in New York at three P.M., it took two minutes to team us up the first time and one minute the next. Staff here never made it past the first round. The first question was, as of May 1, 1990, how many Rocky movies have been made. Unfortunately, another player was either faster or closer to the correct answer.

Had we gotten to level two, we would have been given a secret ID code number and an 800 toll-free number. To win the $1 million, the player must be the first to answer one question correctly in each of seven rounds. The prizes for winning rounds two through six are five dollars up to $10,000. The first $1 million will be awarded May 20 and a new $1 million winner will be named every second Sunday thereafter. *May 1990*

Big Time Games
Jeopardy & Wheel Of Fortune Come to 900

Starting in August, two new telephone games with high name recognition will enter the 900 marketplace.

Players International Inc., a marketer of gaming products from Calabasas, Calif., announced the signing of a licensing agreement with a division of Merv Griffin Enterprises to produce *Jeopardy!* and *Wheel of Fortune*, in the United States, Canada and Puerto Rico.

Call Interactive, a joint venture of AT&T and the American Express Information Services Company will be processing the calls, which can be placed for one to two dollars a minute.

Players who respond within the seven second time limit, remain in the game, and can win daily prizes, said David Fishman, the vice chairman of Players International. *May 1990*

Family Feud Telephone Game

The Family Feud Telephone Game, developed by the Phone TV division of LBS Communications Inc. under an arrangement with Mark Goodson Productions, offers daily prizes of $100 and a grand prize of $5,000 to contest participants. Callers to 900-USA Feud are able to answer questions and, for daily watchers, enter a code and receive bonus points. Service Bureau support is provided by

Audiocast with Telesphere providing the network services. *June 1990*

Game Counseling
Over-the-Phone Help

Just Like Pratt Whitney and Lotus 1-2-3, LucasFilms is turning to 900 for help. The company previously used an 800 number for counseling players of games like Loom and Indiana Jones and the Last Crusade. BFD Productions, a Las Vegas service bureau, created the 900 line for players to get helpful tips with game situations.

The call to 900-740-Jedi is 75 cents a minute and helps LucasFilms make some money from pirated copies that don't have the documentation needed to play the games. *Feb. 1991*

Dial-A-Joke

Dial 900-USA-Danny and you will get sound bites of Vice President Quayle's best known bloopers. The line, a brainchild of Illinois entrepreneur Gary Cohn, costs $2 for the first minute and 99 cents for each additional minute. *Dec. 1990 - Jan. 1991*

Hollywood Squares Goes Interactive
Davidson Hosts Phone Game

Teleline, a provider of voice information services, is making the game show, *Hollywood Squares*, accessible at home 24 hours a day by dialing 900-860-4005. The game follows the *Let's Make a Deal* 900 program that Teleline launched this past February.

The *Hollywood Squares* Telephone Game started on June 15th and is formatted like the television game. John Davidson, the TV show's popular host, duplicates his role posing actual questions from the original game. As Mr. Abbe Goodman, of Howard J. Rubenstein Associates, Inc., the Public Relations firm handling the game, put it: "Hearing that familiar voice really helps people be part of the television game."

Calls cost $2 for the first minute and $1 for each additional minute, and the clock keeps ticking when players successfully complete the first round and subsequent bonus and "super square rounds." Grand prize drawings are held once a month and winners are eligible for the grand prize 1990 Isuzu Impulse as well as other valuable cash, savings bond and merchandise prizes. *July 1990*

Largest Interactive Contest
In 3 Weeks 4.7M Calls Generated

For service bureaus and IPs, planning applications centered on games and contests (see March '91 *Audiotex News*) definitely continue. Brian Rivette, vice president of marketing for Call Interactive, reports a total of 4.7 million calls generated during a three-week promotion of the 900 telephone version of *Wheel of Fortune*.

On November 1, the first day, 300,000 calls were received while more than 400,000 were received during the final day -- November 21.

Calls cost $2 a minute and a portion of the total call proceeds helped provide toys worth $1 million for Toys for Tots.

Pat Sajack introduced a puzzle each day on the television game show, inviting viewers to dial in and instantly solve the puzzle. Players entered the first missing letter to the puzzle by pushing the corresponding key on the telephone. Each caller was eligible to win a Caribbean cruise and daily $1000 cash prizes. *Feb. 1991*

Loews Movie Mania Challenge
Trivia Game Uses IIA Standards of Practice

Starting March 6 and running for eight weeks, Tele-Publishing Inc., a Boston based service bureau, launches Loews Movie Mania Challenge, a telephone trivia game.

Callers test their knowledge (for $.99 per minute) of movie trivia by calling 900-97 MOVIE. They hear a series of questions of varying skill and point value. Each week $500 is given away in each of three regions to the player or players with the highest score for their best three games. At the end of the eight weeks, the player with the highest score nationally wins a VIP trip for two to Hollywood.

A scratch-off game card given out at the box office contains the 900 number and an actual game question. Tele-Publishing believes this strategy will solve the problem of how to get moviegoers to call after they leave the theater. The answer is in the scratch-off, 60 second on-screen trailer and lobby posters, newspaper and radio advertisements.

The game was developed in accordance with the Information Industry Association (IIA) Standards of Practice for Contests and Lotteries. Peter J. Brennan, Chair of the IIA Voice Information Services Division and a consultant on the Loews project, explained: "The intention is to involve the movie-going audience and not to exploit." *April 1991*

Humor Calls

The 900 industry has changed since 1974 when Floyd O'Neil launched Henny Youngman's Dial-A-Joke, a simple barge-in line with the New York Telephone Company.

Andy Scarpati, President of the Comedy Cabaret Inc., a comedy club chain, claims that he has launched the country's first-ever joke and contest line. Using an interactive menu, callers can listen to jokes, listen to ethnic humor, hear comedians recorded live on stage, compete in the weekly joke contest, listen for the weekly password for reduced admissions and prices, and learn about club information and special events. Dial 900-773-JOKE at $1.49 a minute -- you can use a rotary or touch tone. *Aug. 1991*

Direct TV Ads and 900 --Way to Go!

More than 1.6 million ABC soap opera fans responded to *Episodes'* 900-number direct response TV ad, which ran from November through March. Aired twice daily during broadcasts of ABC soap operas *Loving, All My Children, One Life to Live* and *General Hospital,* the 30-second spot built the circulation base while collecting subscription fees up-front.

The call-cost for *Episodes*, a bimonthly devoted to ABC Television Network soap operas, was $3 including the annual subscription fee. *July 1990*

900 Title Trouble for Church Musician

Those who thought the tide of consumer acceptance might be turning should consider the case of Gregory Lunsford, a church musician criticized for his single "1-900-Love Confession." Lunsford said he was forced to institute a refund policy for anyone who thought the song was "nasty."

Lunsford lamented that some radio stations were not playing the song because of the title. *Jan. 1993*

Concert Tour 900 Hotline

Callers in search of information on when and where their favorite musicians are playing can call Star Quest, Inc.'s concert tour hotline. This two-year-old line gives concert dates, cities, venues and telephone numbers for ticket information at $2.99 per minute. Callers can access information by either musician or area. Print ads for the line list three-digit codes for each performer and area.

According to New Hope, Pennsylvania-based Star Quest, the line is designed for speed, minimizing expense to the caller. The service bureau is Scherer's Communications Inc.

Star Quest uses multiple 900 numbers to track advertising efforts. Callers to 900-288-4630 who use the 257 access code can hear the Garth Brooks schedule. *Nov. 1993 (active)*

ABC Uses 900 for "Wild Palms" Help

Viewers confused by the turn of events on the recent four-part ABC miniseries "Wild Palms" could dial 900-773-WILD. The 75 cents-per-minute line offered summaries of what happened so far and previews of upcoming action. *July 1993*

Lawyer Joke Line

There have been law lines and there have been humor lines on 900. Now there's one that's both. The Shark Line at 900-26-SHARK lets callers hear jokes about lawyers for $1.49 per minute.

The line is very tongue-in-cheek, and many callers are lawyers themselves, said Steve Lorenz, president of the line's parent corporation Doantsumi Corp. (pronounced "don't sue me"). "They like poking fun at themselves," Lorenz said.

Lorenz, who runs Doantsumi with his wife, said that the idea for the line came from his work as an insurance agent for contractors. His clients sometimes get sued, a good source of inspiration for lawyer jokes. The service bureau for the Shark Line is Teleshare 900.

When the line started on April 1, it contained jokes recorded live at Hilarities Comedy Club in Cleveland. Subsequent jokes on the line come from calls to Doantsumi's 800 number set up to collect contributions. Callers whose jokes help update the line every 10 days receive a free t-shirt.

According to Lorenz, the line is going well and the free publicity generated from press releases sent to newspapers and interviews on radio stations is a big help. The line gathered 700 calls in its first month of operation.

Lorenz told *AN* that he has only put $300-400 into advertising, including the Village Voice, American Bar Association Journal and a local business paper, and doesn't intend any more advertising. "It's hard to find the best target market," Lorenz said. Instead, he will concentrate on the publicity end, particularly radio interviews. "These tend to spur newspaper articles because someone [in the business] hearing the interview calls you for an interview."

Lorenz anticipates the Shark Line as short-lived, lasting perhaps only three to four months before it runs out of material. "There are only so many lawyer jokes," he said. *June 1994*

Clinton Joke Line

In the humorous spirit of the Dan Quayle Blooper Line, now there is the Bill Clinton Joke Line at 900-990-9601. The $1.50-per-minute line allows callers to hear up to four minutes of jokes about President Bill Clinton.

Callers can also add their own jokes for possible inclusion on the line. The line is sponsored by a Florida resident who wants to use the profits to publish a book of jokes. *Jan. 1994*

Dial-A-Joke Revisited
Henny Still Strong

Audiotex industry historians agree that Dial-A-Joke, which began on 976 back in 1972, was a pioneer pay-per-call application. Now, more than 20 years later, comedian Henny Youngman, who initiated

---

Dial-A-Joke, is turning to 900 in a weekly updated "Henny's Joke Line."

Master of the one-liner, Youngman plans to produce a separate insult line. At 88 he is still performing in nightclubs throughout the U.S. Automated Fax & Voice Solutions is the service bureau for both lines. *May 1994*

Faxed Comics

Who says there's no fun in 900? The Detroit Free Press now offers up to a week's worth of comics via fax through its Free Press Plus. Callers can request such old favorites as Archie, Gasoline Alley and Dick Tracy from 900-740-PLUS. Each call costs $2.50 per minute. Most calls last less than a minute, except when ordering multiple strips. A new set of strips is available every Friday.

The paper reports that Mail Boxes Etc. offers fax usage at a special discount to Free Press Plus users. *Feb. 1993*

Line Gets National Publicity
Reel Review Uses Publicist to Market

Sharon Kissack is an IP who has seen her 900 line get national publicity. A September *USA Today* article featured Kissack's "Reel Review." The line at 900-903-1117 offers brief summaries of current movies with a focus on subject matter, such as violence, which can be offensive to viewers.

In addition to the *USA Today* coverage, "Reel Review" has been the subject of other newspaper articles, such as those in the *Boston Globe* and *LA Daily News*. When *AN* spoke to Kissack she noted that *People* magazine had scheduled a reporter to interview her soon. This media coverage has resulted from the efforts of Joy Scott & Associates (818-782-1323), the publicity firm Kissack has retained. In addition to newspapers, publicity efforts for the line are being directed at magazines, broadcast media and video stores, the last a tie-in to the newsletter Kissack has of "Reel Reviews" for new video releases.

"Reel Review" has been steadily receiving more calls since its beginning in July, with close to 200 calls received in September. The average call length is three minutes. Although she thought the initial call count would be higher, Kissack is encouraged, and believes that the line just needs more media exposure. Scherers Communications, Inc. is the service bureau.

"Reel Review" is designed for use by adults in determining what movies they want to see, and for parents, especially those of teenagers, in discussing movie content with their children. "It's really a service I would have loved to have had when my kids were younger," Kissack said.

Updated weekly, the approximately 40 reviews are non-judgmental, summarizing the movies' content in terms of elements like violence, sex and swearing. Kissack stresses that she does not follow an agenda, such as being anti-violence. "I'm just trying to give people information." *Nov. 1992 (active)*

Music Star's Line

Pop music stars have used 900 lines before, but Michael Bolton is the first to promote his 900 line on an album. The line, at 900-407-BOLTON, offers information on tour dates and Bolton merchandise as well as a message from Bolton at $1.99 per minute.

A portion of the proceeds from the line go to the Michael Bolton Foundation, a non-profit organization to benefit homeless families and underprivileged youth. Wayne, Pennsylvania-based Advanced Telecom Services (215-964-9146) is the line's service bureau.

An associated 800 number allows callers to join Bolton's Gold Club and subscribe to his magazine. *Jan. 1994 (active)*

Preview Line Running Three Years

Since its beginning in 1990, The Preview Line, a movie review line sponsored by the Catholic News Service (CNS), has been cited in the press as a worthwhile 900 number *(AN,* Sept. '92). The line, reached at 900-PREVIEW, has seen a recent decrease in call volumes, but still continues.

According to Thomas Lorsung, CNS director and editor-in-chief, the decrease in call volume seen over the last couple of years is a development which CNS attributes to difficulty in getting the heavy promotion 900 thrives on and the stigma that 900 still suffers from.

The CNS lacks the budget, time or staff to launch a heavy promotion, Lorsung said. Also, space restrictions at the Catholic newspapers the wire service submits material to can leave out the top 10 movie list CNS sends out to promote The Preview Line.

In addition to the top 10 movie list produced in conjunction with Variety, CNS uses a seasonal graphic with the theme "Summertime Movietime" and does some promotion outside the Catholic market themed "You don't have to be Catholic to call 900-PREVIEW."

CNS has seen the line's call counts climb after positive press in national magazines, Lorsung said, citing a U.S. News & World Report article on 900 that mentioned The Preview Line. Lorsung stressed that the line is not only Catholic in content. The reviews callers can hear (for $1.50 for the first minute and 75 cents for each additional minute) express values not restricted to the Catholic faith, Lorsung said. "The reviews look at the moral and aesthetic value of a film."

Wayne, Pennsylvania-based Advanced Telecom Services is the line's service bureau. *Nov. 1993*

Radio Demo

900 lines are all about information delivered via audio, and there's one new line that's a perfect match for the industry. Radio Demo at 900-225-DEMO features a showcase of demos from vendors in the radio industry.

For $1.29 per minute, staff from radio stations can listen to brief demos from music libraries, comedy services, syndicated shows, air talent and jingle, sweeper and sound effects companies. If interested in a certain vendor, they can leave a voice mail message requesting a complete demo or further information.

"These talents need to be heard," said Kimmery Beels of Kimmunications, the line's IP. "We believe we are the first and only audio directory in the industry."

The line was launched November 15. Vendors can also use the 900 number to retrieve responses from their voice mailboxes. Vendors get their demos on the line by calling a corresponding 800 number.

According to Beels, vendors use the 900 number to update their demos because it is faster than 800, where there is a live operator. The 800 will be kept for new vendors to add demos to the line and as a possible promotional tool for repeat vendors, Beels said.

When *AN* spoke to Beels prior to the line's launch, she said that the line had been gathering great publicity including from major trade publications, such as *Billboard*. The marketing effort prior to November's launch concentrated on reaching the vendors, using direct mail and ads in trade publications. After the launch, advertising to reach radio stations began.

The line's service bureau is Scherers Communications. *Dec. 1993 (active)*

Star Trek: TNG Viewer Poll

Paramount Television's five-hour "Star Trek: The Next Generation Viewers Choice Marathon" aired last month, featured five "all-time favorite" episodes selected by viewers who called a 900 number. For 90 cents per minute, viewers cast their votes at 900-288-8735. The 900 poll and marathon were held in conjunction with the final episode of "Star Trek: The Next Generation." *June 1994*

"Telephone Scrabble" on 900

"Telephone Scrabble," a 21-day promotion kicked off in late April, continues the link between games and 900. NBC would not reveal the number of calls received. The 900 telephone promotion offered viewers of NBC's "Scrabble" the chance to win daily cash prizes and compete for a grand prize of $10,000.

Each day the promotion featured a puzzle introduced during the television show. After receiving a head start clue, viewers dialed a 900 number costing 95 cents per minute. They competed by pushing the corresponding touchtone button for the first letter of the puzzle's answer. Additional clues were given in the average three-minute call, and the fewer number of clues a caller needed, the more points scored. Each caller with the highest daily score won $1,000, and the one caller with the highest score for the entire promotion won $10,000.

Skycastle Entertainment, Inc. coordinated the promotion and Omaha-based Call Interactive operated as the service bureau. *June 1993*

VJN Changes
Leaves Sprint and LO-AD
for AT&T & West Interactive

Video Jukebox Network, Inc. (VJN) has selected Omaha, Nebraska-based West Interactive Corporation (800-841-9000) as the service bureau for its Jukebox Network, now known as THE BOX. Formerly with LO-AD Communications Corp., the program was carried on Sprint Telemedia lines. THE BOX is an all viewer programmed network on basic cable television systems that reaches over 33 million households per day.

West Interactive will process THE BOX's more than 600,000 expected monthly requests utilizing AT&T's MultiQuest ® 900 Service. Callers in 33 states, the District of Columbia, Puerto Rico and the United Kingdom dial 900 numbers to reach their local jukeboxes and use a three-digit code to request specific music videos. The cost is 99 cents per video, and requests can be made for two or three videos at a time.

According to the Miami-based VJN, it selected West Interactive partly because of its technological feature, "two-way drop and insert." This allows all voice processing files to be stored at VJN's facility, while the code of the requested video goes to the local jukebox. Previously, a caller's touch tone signal would pass through the service bureau to be routed to the local jukebox where voice processing files were stored.

In addition to the new name, other recent changes to the three-year-old network include a wider mix of video choices and the addition of ongoing artist promotions. *July 1992*

Music Video Request Line

Video Jukebox Network Inc. is still going strong with THE BOX, the five-year-old viewer programmed network on basic cable television systems that allows callers to request music videos via a 900 number call.

THE BOX switched to satellite transmission last August and processes more than 500,000 requests a month. It reaches 20 million households and can be accessed in all 50 states and in the United Kingdom. Paul Sartain, vice president of operations, could not give further details, but said that THE BOX is currently expanding its international reach beyond the U.K.

Callers to THE BOX pay between $1 and $2.50 per call, and through some cable stations, they can request multiple videos at a time. THE BOX utilizes AT&T's MultiQuest® 900 Service. West Interactive is the primary service bureau.

Currently the only interactive capabilities of THE BOX involve taking video requests from callers. However, Sartain said, Video Jukebox Network is investigating taking advantage of the pay-per-call industry's more advanced interactive capabilities, which would allow merchandising and promotion. Sartain gave the example of a caller who requested a country music video also being offered the opportunity to order concert tickets for an artist in that genre.

Until 1992 THE BOX was known as The Jukebox Network *(AN* July '92). That same year also brought the switch to West Interactive and AT&T lines from LO-AD Communications Corp. as the primary service bureau and Sprint Telemedia lines. *June 1994*

U2 Tour Line

Diversion magazine launched a 900 line for news and concert information about the rock band U2. The line will run for the length of the band's U.S. tour. When the line was plugged on MTV, it generated thousands of calls. In addition to hearing the tour information, callers can leave their names and addresses to receive a complimentary copy of *Diversion*.

Calls to 900-288-8284 cost 99 cents per minute. The service bureau for the line is IdealDial.. *Dec. 1992*

Venture with Publications
Music Access Extends Artistic Promotion

Music Access, Inc., a service bureau dedicated to music promotion *(AN* August '91), has teamed up with two major publications to provide audiotex music support for them and further its own mission of increased exposure for independent artists.

Respective 900 numbers listed in the *Chicago Sun-Times* and the magazine *Sassy* offer recorded samples of various artists and information on where recordings can be purchased.

Bar Biszick, executive director of Music Access, said that management at both the *Chicago Sun-Times* and *Sassy* has been extremely cooperative regarding the programs. "I think that they are recognizing that interactive services in general can be used to expand and enhance print subject material," Biszick added.

Callers to the *Chicago Sun Times* Music Access -- at 900-454-4283 -- use four-digit codes to request 1 1/2-minute song excerpts. Calls are 95 cents a minute, and the average call is about 2 1/2 minutes. After listening to excerpts of their choice, callers can hear information on how to purchase recordings locally or by mail order.

"We want our [twice weekly] listings and features to be a valuable extension of print," Biszick said. Music Access foresees the *Chicago Sun-Times* venture as leading to programming on local lines in the Chicago area being given the necessary corporate sponsorship. "It may be possible for us to locate equipment and offer services on local lines providing we attract a sponsor," Biszick said.

In addition to general listings, the program's music selections are intended to tie into the *Sun-Times* coverage of local music events, and its other editorial music features. The first listing in early June tied into the Chicago Blues Festival.

Biszick noted that the call counts during that first week with the tie-in to the festival were double or more than call counts during subsequent weeks of general listings.

The general listings contain music selections encompassing all major genres of music from rock/pop to classical/New Age to experimental. All selections are recommended by specialized music magazines that Music Access is affiliated with. Much of the music carried is not widely distributed through commercial air play or retail sales.

The *Sassy* Music Hotline started up August 1 at 900-737-2779. Similar to the *Chicago Sun-Times* Music Access, it supports editorial features on music, and callers can obtain information on how to purchase recordings. Calls are also 95 cents per minute. *Aug. 1992 (both lines active)*

Video Game Playing Tips on 900

Prompted by an anticipated surge in customer inquiries about its video game titles, Sony Imagesoft, the video game division of Sony Electronic Publishing Co., launched a 900 number last December. This information line for game players at 900-CALL-SONY came online in conjunction with last year's introduction of eight Sony Imagesoft titles for Sega CD, and more than 20 for Sega and Nintendo platforms.

For 95 cents per minute, callers have access to game hints, playing tips and "insider" information. Las Vegas-based B.F.D. Productions (800-444-4BFD) is the line's service bureau. *Jan. 1994 (active)*

A Heaping of Abuse

If you have a tendency toward excessive self-assurance, perhaps bordering on arrogance, and you need an attitude adjustment, call *Dial an Insult* at 1-900-2-INSULT, at $2 the first minute and $1 each additional minute, for a well-deserved heaping of verbal abuse. This program has been around for several years despite the humiliating treatment it doles out to its masochistic customers. *(active)*

Overseas Adult Lines
809 and 10658-0604 Offerings

Probably the chief export of Sao Tome, a tiny island nation off the coast of West Africa, is sex talk. Setting up overseas sex lines gets around 900 call-blocking and 800 presubscription regulations.

Some telephone company officials say they have heard relatively few complaints about international sex lines so far, but are beginning to get worried. Thus far phone companies' response has been: "Your phone, your responsibility." Long distance companies can block overseas calls for consumers, and while that might work for a family with teenagers, it certainly isn't the answer for every home or office.

Bob Spangler, deputy chief of enforcement for the Federal Communications Commission (FCC), has been quoted as saying there's little else that U.S. long distance carriers can do. However, as we have previously seen, the FCC, and the LECs who ultimately bill the users' phones, are complaint-driven.

The foreign phone companies in places such as Suriname, the Netherlands Antilles, Portugal, the Dominican Republic know what's going on and, in some cases, finance the effort. It's not unusual for a phone company overseas to be government-owned. The Sao Tome phone company is jointly owned by a Portuguese company and the Sao Tome government.

An agreement recently reached with CODETEL, the Dominican Republic's phone carrier, allows American companies to deduct from their payment when a customer protests. Other agreements like this may be forthcoming.

A number of US service bureaus work with CODETEL, the Dominican Republic phone company providing 809 services. Among these is Tel-Ad Communications Inc., a Santa, Ana, California-based service bureau. Tel-Ad offers live chat services. Michael Newton, CEO, said that with programs using 1-809, a regular area code terminating in the Dominican Republic, callers are billed on their regular phone bill, because 809 is considered a US area code. According to Newton, the caller pays the regular long distance rates ranging between 60 cents and $1.48 a minute, depending on the day and hour and location of the caller.

Tel-Ad offers live chat. IPs pay a $1,000 sign up fee and get 24 cents a minute and are guaranteed no chargebacks. The expectation is for 10,000 minutes or a $200 monthly minimum fee kicks in. They've been doing this for a few months and their tracking shows that an IP putting in $10,000 a month can expect to net $9,000.

Newton says that IPs need to put in a minimum of $5-10,000 a month or forget this type of application. Newton claims, "IPs can generate the minutes because the price is so much lower than the $3 to $4 dollars a minute callers pay with 900. And there are no chargebacks because callers must pay their long-distance bills."

Another product Tel-Ad will be offering is through a Vancouver, Canada TelCo. This application is similar to calling using a calling card. Callers dial 10658-0604 and then a seven-digit number. Tel-Ad's president John Masso said, "The cost to the caller is the same as a long distance call to Vancouver would be, depending on the time of day and the caller's location. With these 16 digits, and our program, IPs can do adult programming with a 75 day payout." *July 1994*

900 PREVIEW Discontinued

A continuing drop in call counts have cost the 900 industry an established application, 900-PREVIEW, the movie review line sponsored by the Catholic News Service (CNS). The line began in 1990 and gained positive coverage in the general press *(AN*, Nov. 1993).

Call volume decreased over the last couple of years. The final count stood at 19,000 calls generated for the life of the program. "The system just wasn't producing the way we had hoped it would," said Thomas Lorsung, CNS director and editor in chief.

900-PREVIEW featured capsule movie reviews assessing the moral and aesthetic value of films. Calls cost $1.50 for the first minute and 75 cents for each additional minute. The line's service bureau, Wayne, Pennsylvania-based Advanced Telecom Services Inc. (ATS), initiated the line's discontinuance, saying that the number of calls did not justify the cost of updating or storing the information. "Since this line was one of our originals, it is sad to be losing it," said Bob Bentz, ATS director of marketing. "The information provided was excellent."

"We appreciate all the support they've given the system over the years," Lorsung said of the service bureau.

According to Lorsung, the most success 900-PREVIEW had was when the general press picked up the story. That general press coverage included being cited as one of the most useful 900 numbers by *U.S. News & World Report* and a *New York Daily News* story later picked up by the national wire services.

Lorsung told *AN* last year that CNS lacked the budget, time or staff for heavy promotion. It used a Top 10 movie list produced in conjunction with *Variety* and submitted to Catholic newspapers as a main promotional item. Some promotion outside the Catholic press took place as well, themed, "You don't have to be Catholic to call 900-PREVIEW."

"There's definitely a need for the information," Lorsung said. "An experiment we tried in December 1992 on an 800 number established that." An ad for an 800 movie preview line which ran in 300 parish bulletins drew 1,100 calls in one month. *April 1994*

Adult Programs

Any discussion of the pay-per-call industry would be incomplete without addressing adult applications, otherwise known by the media and the general public as "dial-a-porn." This does not include the entertainment-type lines -- like date, fantasy, chat and romance lines -- which may be imaginative or even suggestive, but in fact are quite tame by comparison. Before going any further, however, we have to define exactly what we mean by the term "dial-a-porn."

For our purposes, dial-a-porn will be defined as containing "indecent" language, defined by the FCC as "the description or depiction of sexual or excretory activities or organs in a patently offensive manner as measured by contemporary standards for the telephone medium." Understandably, such a broad definition will encompass many grey areas, and further discussion of what constitutes indecent language is beyond the scope of this section. Nevertheless, a brief overview of the legislative history is in order. Under certain circumstances, indecent language, or dial-a-porn, is allowed on the telephone for commercial purposes between consenting adults.

Legislative & Historical Overview

Beginning in 1984 there were several court cases and appeals relating to dial-a-porn over the telephone. It all began coming to a head in 1988 when Congress tried to ban both obscene and indecent speech by passing The Helms Amendment to section 223(b) of the Communications Act of 1934, the major federal law governing the telecommunications industry. Shortly thereafter, in *Sable Communications v. FCC*, the U.S. Supreme Court held that the Helms Amendment was too broad in limiting indecent speech along with obscene speech, so Congress went back to the drawing board and came out with another amendment in 1989, banning:

"any obscene communication for commercial purposes to any person," and,

"indecent communication for commercial purposes which is available to any person under 18 years of age."

One of the most important objectives of this amendment was to require common carriers to act in good faith and take technically feasible steps to ban obscene telephonic communication and provide access to indecent telephone communication only to subscribers over 18 years of age who affirmatively request access in writing.

In response to this latest legislation by Congress, the Federal Communications Commission (FCC) issued its Report and Order Concerning Indecent Communications by Telephone (FCC Indecency Rules, now the current law), which prohibits indecent communications except under the following conditions:

A. Reverse blocking must be provided by the carrier, where the subscriber must request such indecent services in writing before gaining access to such services, when payment for such services are through the carrier's premium billing services; and,

B. Take one of the following actions to restrict access to persons under 18 years of age:

1. Require payment by credit card before transmission of the message; or

2. Require an authorized access or identification code before transmission of the message; or

3. Scramble the message using any technique that renders the audio unintelligible and incomprehensible to the calling party unless that party uses a descrambler.

Because reverse blocking is required only if the telephone company bills and collects for the service (i.e., premium billing services), and virtually all carriers have elected to deny premium billing services for indecent programs, these new rules resulted in much of the indecent adult programming to migrate over to 800 lines, using credit card billing.

Indecent services are still accessible over some 900 lines, but without the benefit of premium billing services, so that a third party billing system must be used. These bills are often included in the same envelope as the telephone company's monthly bill through contractual arrangements with the various LECs. The two leading alternative billing companies are VRS, Santa Clara, California and ITA, Atlanta, Georgia. Some IPs actually prefer the services of these alternative billers over the phone company's premium billing because of the level of services provided.

Many adult IPs now use 800 lines and credit card billing for purely economic reasons: less lost revenue from chargebacks. According to Jeff White, director of marketing at R.j. Gordon and Company (this is not a typo, the "j" is lower case on purpose), a Los Angeles-based consulting and financial services company with many adult IP clients, "credit card billing is by far the most effective alternative billing method (for adult services). When we first started in the business, credit card chargebacks were 20% or so, and we started implementing different fraud control and data base systems, and we eventually got it down to 3%. Then VISA mandated 1% as the maximum allowable chargeback rate and we had to become very aggressive in our fraud control systems, and now we operate at a rate of about 0.7%."

Why all the controversy?

Early in the game, when there were few rules and no legislation in place to control the industry, adult programming was available to anyone who picked up the telephone, including children. This uncontrolled access is what really caused all the problems. Parents, or the people who paid the phone bill, were getting completely unexpected charges, often for huge sums, for dial-a-porn services. For good reason, these outraged parents complained vociferously, and a lot of people paid attention, as they are wont to do when sex is the topic, and among them were members of the media and the political establishment. A heated debate ensued, resulting in the legislation discussed earlier. Lost in the debate was the fact that these programs were usually quite honest about what they delivered, and that most of the callers were consenting adults, who, for whatever reason, were willing to pay for the services rendered.

Another reason for the fast proliferation of adult programming was the easy money involved. The early adult IPs were raking in a lot of money, owing to the novelty of such services and an apparently unquenchable demand.

Who is calling and why do they call?

While there are not yet any coordinated studies of the demographics of those who call adult 900 and 976 services (callers do not want to answer these questions at the beginning or the end of the call), some anecdotal information exists. Ted McIlvenna, Ph.D., president of the Institute for the Advanced Study of Sexuality, estimates that three quarters of the callers are married men, and they call to talk with someone who is open and "sex positive."

Two types of phone sex services currently exist: pre-recorded messages and live programs, where the caller talks to a "live" person. According to Al Cooper, Ph.D., Director, and William F. Fitzgerald, Ph.D., Associate Clinical Director of the San Jose Marital and Sexuality Centre, "why they call" likely differs for each type. Someone who knows that he is calling a pre-recorded message relies only on the allure of the advertisement he saw, while the callers to live phone sex programs yield data that begin to suggest a profile.

Estimates by Cooper and Fitzgerald suggest that one quarter of the callers make reference to cross-dressing (men wearing women's clothes), and that almost half of them focus on topics related to their being dominant. A logical explanation (which is supported by the prevailing research and by Cooper's and Fitzgerald's clinical experience) for what often leads a man to utilize this type of service is his fear of sharing his fantasies with significant others. The more the other person's opinion matters, the harder it is to reveal oneself. Fantasies figure as the most private of thoughts, thus sharing them requires a very high degree of communication, intimacy and trust.

Cooper and Fitzgerald believe that the research indicates that suppressing exciting thoughts about sex actually results in an increasing cycle of thought, suppression, stronger intrusion of the thought, stronger suppression, et cetera. Specialists in psychosexual therapy have ample clinical experience to illustrate the ineffectiveness of attempting to suppress sexual impulses. Case in point: "Jimmy Swaggart and Jim Bakker, men whose entire lives seem devoted to the suppression of certain exciting thoughts, became ravaged by the very excitement they had hoped to suppress," asserts Daniel M. Wegner, Ph.D., et al, in the *Journal of Personality and Social Psychology* (March 1990). In most cases, the therapy of choice is substituting healthy sexual outlets for dysfunctional or harmful outlets.

The expert consensus is that live phone sex services are utilized predominantly by men currently in relationships, who are afraid to share their fantasies with the most emotionally important people in their lives.

The telephone provides a "solution" for physical intimacy. The medium facilitates surrogate sex through the *illusion* of intimacy. It is precisely this surrogate function of the telephone which became the essence of the "reach out and touch someone" advertising campaign. The psychological assumption was that a telephonic relationship was a substitute, a proxy, for actually seeing and even touching the person at the other end of the connection.

In-Depth Profile:
Paramount National Media, Inc.

Glen Chazak, owner and president of Paramount National Media, Inc. in Beverly Hills, California (310-282-8317) has been an IP in the adult pay-per-call business since it began in the mid 1980s on local 976 numbers. He now spends his time and resources between both adult and psychic programs, and he has done quite well with both applications. Nonetheless, this profile deals with the adult side of his business.

Adult pay-per-call started back in 1985 on local 976 numbers. In the beginning, virtually all of the adult programs were recorded, with few live operators. In those early days, it was not unusual for the larger adult IPs to make a 20 to 1 return on the advertising investment for their programs, probably a result of the novelty of such programs and a pent-up demand for this type of entertainment. These early pioneering adult IPs, including Glen Chazak, were making huge sums of money. The most successful of these IPs were investing large sums in national advertising in expensive publications such as *Penthouse* and *Hustler*. At a 20 to 1 return, a $10,000 investment in one display ad would return $200,000 or more in revenue.

In 1989 long distance 900 pay-per-call became available to adult IPs, and by then the advertising return ratio was still a very impressive 15 to 1 or so. Along with the advent of 900 came live adult programs, which resulted in longer call hold times and higher per-minute charges as compared to recorded programs. According to Chazak, "the vast majority of the money being made in 900 was from adult programs... and the bulk of the callers, maybe 95%, were male."

Nonetheless, the gravy train was relatively short-lived. After 1990, returns started dropping to more realistic levels as the novelty wore off, chargebacks became an increasingly serious problem, competition increased and people began blocking access to 900 numbers. "By 1992, returns were down to 5 to 1, and then maybe 3 to 1 by 1993," says Chazak.

Now, a 20% to 50% return on monthly advertising expenditures in more the norm. (i.e., 1.2 to 1 vs. 20 to 1). Although this is still a very respectable return by any other standards, "back in 1989, you'd think a 20% or 50% monthly return on your money was disastrous," says Chazak. If you're lucky and you really know what you're doing, it is still possible to earn up to a 2.5 to 1 return on your money, Chazak asserts, something he is able to accomplish because of his long-term experience in this industry.

To make money today in this business, you have to hook up with a pro, says Chazak, who knows how to advertise the program the most effectively. For example, a full page ad in *Hustler, Oui, Club* or *Penthouse* might cost $15 or $20 thousand, but the right one-third page ad can pull just as well for a fraction of the cost. The ad layout, copy and graphics -- particularly the "picture" of the beautiful woman -- are extremely critical to the effectiveness of the ad.

Another change in the business is that a single 900 number in the ad doesn't work any more. The IP needs to advertise several numbers (usually connecting to different women or types of programs, or different "fantasies") in order to give the caller the perception of having many choices. Nowadays, for example, a one-third page ad should have 2 or 3 numbers. "On a full page ad, we will often display up to 9 different numbers," says Chazak. All of his programs are now live, with recorded programs a thing of the past.

Another important marketing technique is to "tag" another 900 number or a 011 international number onto the program, which simply means you are promoting that second number to the caller in the hopes of generating another call. Chazak also feels that it's important to offer not only 900 numbers in the ads, but also an 800-number credit card option as well as a 011 international option. Obviously, giving the caller several choices works very well in this business.

According to Chazak, "you used to be able to spend your last $5 or $10 grand and make a killing. Now it's a business. You better be a good advertising person. You need 5 months of advertising money before you get your first check." At least for monthly mens magazines, with long lead times for placing an ad. "The advertising doesn't hit for two or three months, then there is a 30 day period when the program is called, and then the next 30 days the money is being collected, so it takes at least 5 months to see any money coming back."

For a faster payout, Chazak mixes his advertising between the glossy men's magazines and the weekly newspapers, which can be either straight men's newspapers or the so-called alternative newsweeklies. The lead time for placing ads in the weeklies is much shorter, and the waiting time for getting your money back is quicker. Nevertheless, a glossy men's magazine has a long "shelf life," generating calls for as much as a year after it is published.

Chazak has advertised his programs in virtually every adult publication out there, and feels that the return rate holds up across the board. If a $1600 ad in *Jugs* returns $3200, a $16,000 ad in *Penthouse* will return $32,000. But don't expect these returns at the start. "If you break even on your first month you've got a home run number, because the business is based on residual calls," says Chazak. "The money is made from repeat callers." It will take several consecutive months of repeat advertising before anything like a 50% return will be achieved.

Chazak has used many different service bureaus for his programs, and currently uses more than one. He avoids service bureaus that have their own in-house programs that would compete with his programs, so he sticks with bureaus that provide equipment and call handling services only.

Because of his track record at generating such impressive and consistent returns on advertising, Chazak is now talking with investors and brokerage companies who are quite interested in not only the high potential returns, but also the relative stability and predictability of the returns. Chazak has been outperforming stocks, bonds and other investments by a huge margin on a consistent basis.

Who are the actual female operators in this business? They are women of all ages, shapes and sizes who enjoy earning money by talking on the telephone. There is no stereotype or typical profile of an operator for an adult program. Obviously, they are not particularly shy about talking about sex with complete strangers! They can work out of a phone center or room with numerous other operators, or they can be on a schedule to answer the phone at home. Both methods are used successfully.

The quality of the operators can make a big difference with regard to the level of success for the program. A good operator can keep the caller engaged on the line longer, and will be able to earn a repeat call later. "I could run two identical ads side-by-side," continues Chazak, "with one connected to one set of operators and the other connected to another set of operators. The exact same ad, depending on which set of operators it's connected to, could bring in 5 times as much money without laying out an extra nickel. We have spent years developing relationships and looking for the best operators in the business."

Nowadays, IPs no longer employ their own group of operators. It's simply too much trouble because it becomes a 24-hour operation to manage. The IP now contracts with a separate company that employs a group of operators, essentially leasing their services on a per-minute basis. This means that it's possible to reach the same operator through dozens of different programs marketed by several different IPs. Chazak estimates that there are probably 15 to 20 separate groups of operators, and of these, only 2 or 3 top-notch operations with the well-trained and closely-supervised operators.

When an phone-room operator gets a call, the computer screen tells her what type of call is coming in: an 800 credit card call with no restrictions, or a 900-number call to a "romance" or "chat" line, for example. This alerts her as to what can be talked about over the phone, and the operator is generally given detailed guidelines on what the boundaries are on the two different types of calls.

Chazak's advice to anyone starting in this business is to first team up with an established expert for at least a year. Learn everything you possibly can during that year. Otherwise, "a layman doesn't have a prayer of making it on his own," says Chazak. Also, it just isn't necessary to reinvent the wheel by going it alone and wasting money on ineffective advertising. "There is no reason to experiment." A beginner should team up with an IP, someone who puts money on the table and risks it all. Only an IP with experience in this business has any idea of how to make money. Yes, it sounds easy, but the people making the really impressive money have been in this business for a long time.

When asked where he thought the adult pay-per-call business was headed, Chazak was quite confident that it would be around indefinitely due to the strong, sustained demand for such services. Also, after a brief excursion into 800 pre-subscription, calling card and other methods designed to circumvent 900 numbers, he feels that adult is returning to 900 numbers. This would include the non-indecent chat and romance lines, that have always been okay on 900 lines. The no-holds-barred XXX dial-a-porn will stay predominantly on 800 lines with credit card billing.

The reason for this is because the telephone companies and the regulators are tired of getting complaints from consumers who are confused when they have to pay for an 800-number call. The 900-number industry is already carefully regulated by the FCC and the FTC, and all the necessary rules have been in place since 1993. 900 numbers are a safe place to be in this environment, because there is no need for the carriers or the government to change the rules -- and shut down access to your program -- right before your $50,000 in advertising hits the streets. Because of this, Chazak is now using more 900 lines on his programs, devoting most of his advertising investment to 900-number access to his programs.

Where are we now?

There are those who feel strongly that we have reached a point in time when we may all wash our hands of dial-a-porn, that it's a phenomenon which has long been a messy, unpleasant millstone around the necks of information providers who would rather not be even remotely associated with it. We should, however, acknowledge that dial-a-porn may have well-served the overall information industry by being the subject of a great deal of publicity about 900-number services. Because of this exposure, virtually everyone knows what 900 numbers are all about. The perception may not always be positive, but that is changing now that clear rules are in place.

As we go to press, there are three general ways to access adult programs. Besides 900 (with alternative billing) and 800 (with credit card billing), we're now seeing international dial-a-porn using overseas adult services on lines with 011or 809 (Caribbean) prefixes. These lines terminate in several foreign countries, including Israel, Holland, Hong Kong, Dominican Republic, Sao Tome, Moldavia and a few others. These lines range from tame chat lines to adult sex lines, and present the same problem that plagued early 900 dial-a-porn: uncontrolled access by minors.

It is not unreasonable to expect that other options will come and go in the future, but it will be interesting to see what the smart money is doing. Glen Chazak is going back to using more 900 lines because he's afraid of getting burned on the unregulated alternatives, specifically the international options. This is uncharted territory that can change overnight with new rules, problems or restrictions. It's safer to stay with the proven products: 800 and 900. It's perfectly legal to deliver adult entertainment over these lines as long as the rules are followed, so why indeed risk money on an alternative?

Chapter 11

Chapter 12
Product & Business Promotion & Marketing

In many of the following profiles you will find elements of entertainment and perhaps wonder why they were not listed in the preceding chapter. Well, any good TV commercial must have entertainment value just to get your attention, and a good 900-number program is no exception.

Nonetheless, the major purpose of the following programs is to promote a product, service, event or company; and the entertainment element is secondary to the main objective.

Contests and sweepstakes are other ways to grab the attention of consumers in promoting a product or service, and they have been used extensively in this medium. Unfortunately, some past scams involved sweepstakes lines that never delivered anything of value. Recent legislation has improved this situation considerably, and most such programs that run today are quite legitimate. There are now in place strict standards imposed by the long distance carriers for incorporating contests or sweepstakes into a 900-number promotion.

FanStand Catalog
An increasingly popular way of getting catalogs to interested consumers is through 900 technology. It is an efficient way to bill for catalogs and to develop qualified lists. Consumers can be reimbursed

the cost of the call by using a coupon redeemable with the first purchase.

American FanStand is a distributor of officially licensed sports items and will be marketing their catalog using the Catalog Request 900 Service (900-933-FANS; $3 a call). Their target is new customers. Starting in Albany and Hartford, where American FanStand has retail locations, the company will be advertising on ESPN's Sunday Night Football games.

FanStand, with the help of Advanced Telecom Services, a Wayne, Pennsylvania service bureau, has really embraced 900, and in addition to the catalog line, they are doing another. This line gives real time reports on the latest from the sports world (900-773-FANS; 99 cents a minute). *Nov. 1991*

Chivas & 900 & Sinatra

As Frank Sinatra travels the United States on his Diamond Jubilee World Tour, a 900 number accompanies him. Full page ads for the show placed by local sponsors and Chivas Regal invite callers to win an evening of music with Sinatra.

Net proceeds go to Variety Clubs International Charities. The cost of the call to 900-420-CHIVAS is $2 a minute with a maximum of 2 minutes. Callers who are 21 years or older have the chance to tell Frank they love him, and to win tickets to the show and a CD set of Sinatra's The Capitol Years. *Dec. 1991*

900-Revlon No More

According to a Revlon spokesperson, "there are no plans to use a 900 promotion" to introduce its No Sweat antiperspirant and Revlon Internationals shampoo. There had been industry speculation that Revlon would repeat their Revlon 900-campaign of mid-February as part of a $2 million beauty-care sweepstakes that targeted 52 million households.

Regarding the result of what Revlon believes was a pioneering effect in the 900 interactive applications, the spokesperson said, "they are not disappointed with the results. 900 is not going to be used for other promotions because with a change in management there had been a change in focus." Tom Goeler, the marketing vice president who had promoted this approach, is no longer with the firm.

As part of this promotion, Revlon will give a portion of each call to charity and had anticipated generating about $50,000. The exact amount has not been calculated because final figures are still not available. In capturing callers' names and addresses, Revlon was pleased with the service Bellatrix Communication, Inc. of Denville, NJ provided as the inbound telemarketing service bureau. *May 1990*

Beer-800-Summer-Sweepstakes

Look for a Memorial Day/ Summer sweepstakes and database building beer promotion developed by Modem Media, an ad agency specializing in voice response 800 and 900 programs.

Coupled with heavy TV support, an 800 number will be placed on hundreds of millions of cans with thousands of instant prizes. *May 1991*

ValuePhone & Co-Op
Promotions Make It To Carol Wright

After making a 900 phone call, consumers last month began receiving rebate and general offer certificates. Cost per call is $1.50 for the first minute and 95 cents for each additional minute.

ValuePhone, a consumer promotion developed by Kreissman Marketing and Donnelly Marketing of New York, is making the number available through "Carol Wright" envelopes. Those envelopes represent a direct-response mailing that reaches 30 million households monthly. Consumers find the offers on the inside of the outside envelope (no wasted space here).

After consumers dial 900-454-VALU, they give their name, address and phone number, then listen to the list of offers, press the code number for the first offer they want and a code number for each additional offer.

With an average $4-6 call, certificates worth $10-100 can be gotten with as many offers as are available.

Call Interactive is handling the calls and T-Com is doing the laser imaging and mailing. Advertisers pay $1500 to become part of the value line promotion plus 50 cents for each certificate sent.

Kreissman Marketing has been working on this for some time and is watching the results closely. "We believe this is a breakthrough in both co-op promotions and 900 marketing," said Gary Kreissman, the president. "We married these two ideas, offering both consumers and marketers a tremendous return on investment." *March 1991*

After Eight 900 Promo

Congratulations to After Eight, not only for making my favorite chocolate thin mints, but for embracing 900 technology. With a self-liquidating promotion, they further the image they have been trying to promote: buying and serving After Eight mints is the essence of good manners.

It's good business to get your name before the public, develop a mailing list, if so desired, from the name and/or number-capture and be able to advertise your 900-number on the packaging.

For 75 cents, callers to 900-246-AFTER8 can choose from four updated, monthly categories of etiquette tips. They can choose to leave a question and, if it is used, win a year's supply of mints. *April 1991*

Teleline & Showtime
Handle Pretty Woman Sweepstakes

Teleline, Inc., has entered into a licensing agreement with Showtime Networks to create and operate an interactive 900 number sweepstakes teleprogram to promote Showtime's upcoming broadcast of *Pretty Woman.*

Showtime, the cable network, is furnishing on-air promotion support. And when 11 million customers open their bills they will find additional promo for the 95-cent call.

The grand prize will be a three-day, $20,000 shopping spree in Beverly Hills, a round trip ticket to Los Angeles, a stay at the Beverly Hills hotel and a chauffeured limousine. Five hundred dollars a day is added for miscellaneous expenses. At the conclusion of the 12-week sweepstakes, a winner is chosen randomly.

For 50 instant winners randomly selected and notified while on the call, there are $100 prizes, and another 5,000 callers get audio cassettes. *April 1991*

Halloween Treat

In 125,000 grocery and liquor stores nationwide, a life-size display of Elvira will announce the new Coors Light Elvira Haunted Hotline, 900-680-0707. Priced at $12.95, callers hear Elvira "Mistress of the Dark" give ghostly party pointers on how to have a successful Halloween party. In addition, each caller receives a life-size stand up Elvira, that is identical to those displayed in the stores.

IdealDial of Denver designed and operates the line. *Oct. 1991*

Readers Sweeps and Raps

Bantam Books Loveswept novels will carry details about the "Vive La Romance" sweepstakes during October through December, when winners will receive a two-week vacation in Paris with hotel accommodations. Paired with the contest is an opportunity for the readers to hear an exclusive message from their favorite romance novelists by calling 900-896-2505 for 95 cents a minute. In addition, from 2:00-4:00 p.m. every Wednesday, callers are to be chosen at random to speak directly to a Loveswept author. *Sept. 1990*

Popeye's 900 Value

The *Washington Times* reported that Janjer Enterprises Inc., which operates 18 Popeyes Famous Fried Chicken stores in the Washington area,, is busily promoting its new "Popeyes Value Club." Customers call a 900 number and get $25 in food coupons about six days later. The phone call is $2 a minute, and the typical call lasts two minutes. The campaign is a joint venture between Janjer and Marketing & Research Resources of Silver Spring, Md. *Dec. 1991*

Gatorade Sweepstake

The Quaker Oats Company, maker of Gatorade® Thirst Quenchers, and Call Interactive have teamed up with NFL players in the sports beverage's first interactive telephone promotion. The promotion is one of the first for Call Interactive Direct, a division of Call Interactive devoted solely to direct marketing campaigns.

"Play With the Pros," a national sweepstakes offering hundreds of prizes, began last month and will continue to the end of this month. The grand prize winner and 10 of his or her friends will play in the ultimate flag football game in conjunction with the NFL Pro Bowl on February 2. The group will receive a five-day trip to Hawaii where they will join seven present and former NFL players for the flag football game. They will also attend the Pro Bowl.

A recorded message by Phil Simms, New York Giants quarterback, greets callers. Simms delivers the sweepstakes "game plan" by instructing the caller to enter a two-digit extension number found on "Play with the Pros" promotion material, as well as the caller's name, address and home phone number.

Callers are automatically entered into the sweepstakes, with winners randomly selected the first week in November. Cost is 95 cents per minute, with the average call lasting two minutes. Consumers can also participate via mail entry. *Oct. 1991*

Litigation Against IP Started
Unauthorized 900 Number
For Shell-Shocked Turtles

Surge Licensing, Inc. holders of the exclusive world wide licensing and marketing rights to the Teenage Mutant Turtles, filed suit against SBK Records, Phone Programs, Inc. and AT&T. SBK is running a 900 application featuring the Teenage Mutant Turtles. Since June Surge has been unsuccessful in its attempts to get an injunction to stop the allegedly unauthorized 900 number.

Frank Funaro, a spokesman for Surge, said the suit was filed after the alleged offenders didn't respond to a request to shut the phone line off voluntarily. Surge Licensing is seeking compensatory and punitive damages against the defendants. If the suit is successful, funds awarded would be earmarked for charity and to reimburse callers, according to Mark Freedman, the president of Surge Licensing.

Lawrence Katz, an SBK vice-president for business affairs, said the suit has no merit; AT&T said the phone company only provides the phone lines. *Aug. 1990*

Direct Mail and 900
Incredible Potential To Be Fulfilled

"900 numbers will be a $2 billion business by the end of 1991," said Stan Rapp in the kick-off session of the first Audiotex Conference specifically focused for the direct-mail markets. The conference was sponsored by the Maxi-Marketing Institute at the Plaza Hotel in New York.

Mr. Rapp clearly feels that direct-order has been slow to pick up on 900. The president of the CCR Consulting Group and a director of the MaxiMarketing Institute as well as the co-author of the widely read book MaxiMarketing, had many insights and ideas about audiotex. Rapp was extremely generous in throwing out beginning ideas to stimulate IPs and Catalog Houses. Here's a sampling from his fertile mind.

Catalog Readership Games

Lucky numbers can be distributed throughout the pages of the catalog. Readers dial 900, punch in the lucky number with their personal identification code (PIN) number. The more times you enter, the greater the discount.

Private Sales Prices Numbers

Calls yield special prices for the sale-minded shopper, items can be moved quickly with reductions, and when they are no longer available, recorded messages can suggest alternatives.

Dutch Sales Numbers

With this type of sale, prices for items go down each day until they are sold out. Customers can order using their PIN and credit card.

Happy Birthday Numbers

For marketers of children's merchandise, the caller's first name is inserted by computer programming or digitized sound to have the child listen to his or her very special birthday greeting. This identifies special groups and develops valuable marketing lists.

Guess The Best-Selling Item Game

Each caller gets three guesses to determine the three best-selling items. Each day the first ten callers with correct answers get their choice of any three items free.

900 Catalog Offerings

This helps defray the cost of a full-color catalog for which marketers need payment. A solution when customers are unable to use charge cards for a $3-$5 cost and are reluctant to write small checks, this assures payment and simplifies the ordering process.

Combining a "missed you" mailing, 900 and a special incentive gift, inactive customers can be revived.

Rapp as you can see is full of ideas and is sold on 900 as a marketing tool. "Not for a moment" does he believe that pornography is the real reason why some companies are not quickly going for this marketing tool. He feels "it's a lame excuse for resisting innovation. It's the same kind of resistance encountered when video tapes first entered the scene. Now there is a store on every corner and porno tapes are a dying part of the video industry. It's a phony issue and the new technology offers enormous potential, which will be fulfilled. The question is, who will do it first, and make the most of it." *July 1990*

Quaker Oats Pushes 800/900

Cycle Dog Food, a Quaker Oats company, is sponsoring an interesting 800/900 promotion. PetLine is a 24 hour telephone advice and information service for pet owners. On the 800 numbers, callers leave their name and address, which will no doubt prove a valuable marketing tool, and receive a free directory listing PetLine's topics. Then, for 95 cents a minute, callers can access any of 300 topics by dialing 900-990-PETS.

A caller dialing the 800 number is told about and given the new 900 number, but the consumer is not advised of the cost for dialing 900.

Audiotex News spoke to Diane Primo of Quaker Oats, who said this was an oversight; thanked us and said it would be corrected. The January promotion, which features ads for the 900 number on bags and cans of dog food, did have the cost of the call. *Feb. 1991 (active)*

Doubleday Attempts 900
First Test of its Kind

The Doubleday Literary Guild, a 500,000-member book club, launched a 900 program that helps members decide which books to order by permitting them to hear authors read from their novels.

The message is changed every 21 days to coincide with new promotions. Using the Guild's revised membership magazine, three separate control groups are being used to test the Home PreviewLine.

The tests measure three 100,000 member control groups using either the 900 number, an 800 number or a customer-service number. Kenneth Cooper, director of advertising and creative services for Doubleday Book & Music Clubs Inc., which created the promotion, said the 900 number is being promoted exclusively in the magazine as "an easy way to shop in the comfort of your home."

Cooper told *AN*, "I am enthusiastic about 900 and see it as a fortunate opportunity for new growth for savvy marketeers."

Cooper wouldn't disclose initial response rates from the control groups but noted that the 900 number's conversion-to-sales rate averaged 10-15 percent during the first 21 days of the program. "We wanted to develop an impulse-buying zone, that might not have been generated with the magazine," he added.

Call Interactive, Omaha, Neb., is handling all the inbound calls. The book club is "hoping the ... 800 number generates enough incremental sales that we can afford to offer the service as a toll-free number and won't have to charge our members anything," Cooper added. "The Guild decided to test the 900 number at the lowest conceivable price of 50 cents a minute because we just wanted to break even on costs."

If the service proves successful the Guild will use it in a variety of ways, such as cross promotions with other Doubleday book clubs, gift programs and offerings of discounts on products.

Dialers of the Home PreviewLine are given a menu of options, including which of six authors to call. *AN* listened, for example, to Danielle Steel describe Heartbreak and to Ira Levin read the opening lines of Sliver.

"If a caller listened to every author, it would take 11 to 14 minutes depending on the cycle," Cooper said. *May 1991*

Catalogs and Pay-Per-Call

According to the June *Target Marketing*, catalogs and 900 numbers make a good team when used in proper context.

A positive example they cite is Critics' Choice Video, a cataloger based in Elk Grove Village, Illinois. Last July they launched a 900 number program called Video Search Line, a service that helps customers locate hard-to-find titles as well as titles not listed in the quarterly catalogs. The charge for the call is $1.95 for the first minute and 95 cents for each additional minute. Catalog customers who buy a tape through the service receive a $5 discount certificate toward their next purchase. *Sept. 1991*

Ralston Purina Puts Bread On Table
Second Harvest Gets $5 Phone Call

Fortune 500 companies are increasingly looking to 900 as a way of promoting their products or company images. Look for a story to run soon in *Fortune* magazine that researched (with *Audiotex News'* help) this subject.

Ralston Purina is in that group with a number of applications usually having a not-for-profit partner. A recent example is Home Pride Bread, a product of Continental Baking Company, a subsidiary of Ralston Purina that sponsored a Hunger Information Line (900-468-GIVE). Callers had an opportunity to donate $5 to Second Harvest, a private agency that provides food for 184 national food banks and some 40,000 charities.

During the months of November and December, Home Pride Bread took 5 cents from every loaf sold and used the $100,000 raised to sponsor the info line. Kerry Lyman, in marketing communications for Ralston Purina, said, "The 900 number was on the bread package but we also did extensive news releases and ran full page ads for Second Harvest. The final counts are not in because despite the fact that the line was scheduled to close the end of December, we will keep it going until the calls stop."

Everyone connected with this project was happy --what a valuable and timely Thanksgiving and Christmas promotion! *Feb. 1991*

Increased Postage Helps 900

Increased mailing costs may turn out to be a boost for 900. Direct marketers concerned over rising U.S. postal costs are asking all mailers to search out other media for identifying real prospects and alternative response streams. The obvious answer is getting 900 to work for them. 900 can increase response, lower cost per lead per sale, as well as develop customer data and opinions. *June 1990*

900 to Promote National Tour

Cooking Light magazine is using a 900 line to promote the "Ask Cooking Light" 1993 National Tour of a mobile cooking school and test kitchen. Callers to 900-773-LIGHT hear times and locations for the magazine's 40-foot truck touring the country and get an opportunity to receive $15 in discount coupons.

The decision to use 900 was based on the desire to make tour information and discount coupons widely available while partially offsetting costs, said Kent Back, marketing director of National Consensus, the company that developed the tour promotion. The 900 number is advertised in Cooking Light and People magazines.

West Interactive is the service bureau. Calls cost $1.95 for the first minute and 95 cents for each additional minute.

Sponsored by 15 major companies, the tour will continue until the end of October. Visitors walk through the truck, test foods and receive gifts, recipes and coupons. *Aug. 1993*

American Baby Mag Contest

Thousands of parents competed for instant prizes plus a grand prize of a $25,000 college scholarship, during the *American Baby* phone promotion held in connection with the magazine's TV special, "Discovering the First Year of Life."

Following the special, which aired on the Family Channel twice in March and twice in May, callers dialed 900-786-BABY at 99 cents per call. Each caller had to give a "mystery word" revealed during the broadcast in order to have a chance at a prize. Phone lines remained open for 48 hours after each airing, and according to Phone Programs USA Inc., the service bureau, thousands of calls were received each time.

Entry forms and contest rules were also available in *American Baby* magazine. *July 1992*

Draft Team Hotline

The Hitch Hotline for the cross-country tour of Reminisce magazine's Belgian Six-Hitch Draft team gives callers weekly updates on the draft team's progress. Calls to 900-88-HITCH cost $1 per minute and are approximately two minutes long.

The weekly messages give details on the draft team's activities, rest breaks and next destinations. As of early August, the team had traveled more than 800 miles.

The 900 line is similar to last year's Lewis and Clark Line *(AN* Sept. '92), which gave updates on an expedition retracing Lewis and Clark's historic journey. *Sept. 1993*

Free Books through Silhouette's 900 Line

Silhouette Books has launched a 900 line for callers to hear excerpts from its new and upcoming releases. A call to 900-288-2-BKS will cost $2.

Callers leave their names and addresses to receive two complimentary novels, shipped to them in four to six weeks. *April 1993*

Space 900

Now 900 use is really going places. As a promotion for Arnold Schwarzenegger's next film, *The Last Action Hero*, Paramount Pictures rented space on a NASA rocket, enough for a 50-foot billboard and a compact disc holding taped messages for any aliens in space. Prior to the May 1 launch date, humans interested in leaving messages can call 900-9-ROCKET. Each call costs $3.50. *May 1993*

Get Relief Fast

Whitehall Laboratories, manufacturers of Anacin®, offers a stress relief line with tips on how to manage stress in this hectic day and age. Call 1-900-329-STRESS at $1.50 the first minute, 75 cents each additional minute, and you can also receive a helpful brochure and samples of Anacin.

TBS Uses 900 for "Sweepstakes"

TBS Superstation used 900 in its "Win a Hog!" sweepstakes run in conjunction with the television special, *Harley-Davidson: The American Motorcycle.* The special featured one promotional ad letting viewers know that, through a 900 number or a postcard, they could enter the sweepstakes to win a motorcycle.

Each call to the 900 line cost 99 cents. The special aired twice in March; its first airing drew 2,262,000 households as viewers. In total, 10.7 percent of the households viewing dialed the 900 number. Omaha-based West Interactive Corporation, the service bureau for the number, processed more than 400,000 calls during a five-day period.

500,000 postcard entries to the sweepstakes were received by March 30, for a combined response rate of 24 percent. *May 1993*

Chapter 12

Chapter 13
Fundraising & Charity

A growing trend in the pay-per-call industry has been the use of 900 numbers for fundraising purposes. These programs are often flat rate calls, and usually include a recorded message regarding the topic of concern, often by a recognized celebrity who champions the particular cause. Some organizations that have already experimented with 900 fundraising include World Vision, Leukemia Society of America, March of Dimes, Amnesty International, and Mothers Against Drunk Driving, to name only a handful.

What makes 900 fundraising so promising is the convenience, to both the donor and the fundraiser. It's much easier to dial a phone number than to write a check and address an envelope. The fundraisers' collection costs go way down, and the problem with generous pledges who never come through with the cash is completely eliminated.

Sweepstakes Donates Proceeds to Homeless

A two-minute spot at the beginning of the video release of the motion picture City Slickers, starring Billy Crystal, will promote a "Wild Vacation Sweepstakes."

Cost of the call is $1.95 per minute with a four minute call average. The sweeps will run until June 30, 1992, and all proceeds will go to Comic Relief, an organization dedicated to helping the homeless. West Interactive Corp. is the service bureau and the

Creative Services Group developed the promotion which also features stand-up cards in the video stores.

Save the Dolphins

If you're concerned about the welfare of our dolphin friends, call the Dolphin Project's 900 line, 1-900-USA-DOLPHIN, and the proceeds ($5 the first minute, 50 cents each additional minute) support its efforts in protecting these lovable creatures.

Help War Victims

You can help the victims of the war in Bosnia-Herzegovina with a $14.95 donation to the International Relief Committee by calling 1-900-40-PEACE.

Americom Boosts 900s
Rubber Duckies Help Charity

Americom's CEO, Greg Wood, is coordinating, at no charge, a 900 charitable promotion to help the San Francisco Boys & Girls Club and to improve the public image of the 900 industry.

Joining with him are LO-AD and Telesphere who have truly reduced operating costs. Each $5 call to sponsor a duck yields $4.64 for the charity.

Individuals and corporations are adopting the 50,000 rubber ducks which will be released on October 20, 1990 to run the rapids (courtesy of S. F. Fire Department hoses).

The national number is 900-321-DUCK. Any promotion will be greatly appreciated. *Sept. 1990*

Foundation Fundraising
$50,000 For Coming Up With Right Word

The National Cristina Foundation (NCF) is searching for a new word or a combination of existing words that express the accomplishments of the disabled. Anyone coming up with such a name has the chance to win $50,000.

A 900 number for additional information and literature regarding the contest is 900-WORD. Calls cost $3 each, proceeds helping NCF.

Entries must be submitted prior to November 30, separately, either in an envelope or postcard, to: NCF Contest, 2301 Argonne Drive, Baltimore, Md. 21218. *Nov. 1990*

900 Number for Hunger Relief Starts

Look to hear about a 900 fundraising number for United State, Interaction, an umbrella group representing more than 100 hunger relief organizations. *June 1991*

Fundraising Goes 900

In an effort to raise money for public programming, WGBH, Channel 2 of Boston, Massachusetts, used a 900 number, 900 246-WGBH, this past June for viewers to play The Auction Game. Trans National Group Services, also of Boston, provided the service. A total of 10,235 calls came in. Call-cost was $2.50 per minute. The total number of minutes was 42,342, each call averaging 4.1 minutes.

Trans National says that the system takes minutes to test the media's effectiveness. For example, 75% of the response to a television spot featuring a 900 number will occur in the first fifteen minutes after the spot runs. There were two promotional spots per hour each day of The Auction Game. The average response rate per spot (number of calls divided by program rating each time the spot aired) was 1.33%.

WGBH's 900 number had a taped voice which allowed callers to guess the price of six donated gifts. Placed on a table were the gifts, including vouchers for cars and trips. At the end of the daily game, the caller whose estimate was the closest to the gifts' value won all the prizes on the table. The game lasted eight days with a new table and six new prizes each day. The total retail value of the prizes for the eight days was $20,948. The net WGBH revenue was $59,348. *Dec. 1990 - Jan. 1991*

Mandela Freedom Fund

Contributions may be made to the non-profit Mandela Freedom Fund by placing a $5.95 call. Callers to 900-230-8880 hear a personal message from Nelson Mandela, president of the African National Congress. *March 1992*

ATC Waives Fees for AIDS Program

Automated Tel-Com (ATC), out of Springfield, Mo., is underwriting a new fundraising program for donors to the American Social Health Association. Calls to 900-407 AIDS are $5. ATC has waived all customary fees. *April 1992*

$10 Call Wins Popularity Contest

WNET-TV Channel 13, a public service station in New York, reports that during a six-day pledge drive, when it raised $250,000 in pledges from 50,000 calls, the $10 line was the most popular. Contributors had the option of calling different 900 lines for $5, $10 and $20 pledges. *April 1992*

Adrift But In Touch

Richard Wilson hopes to break the world record for clipper ships sailing from San Francisco to Boston. The trip will take about two and one half months during which time calling 900-820-BOAT provides a three-minute recorded update for $3.50. Some of the proceeds will go to the American Lung Association and the Student Ocean program. *Nov. 1990*

900 And Not-For-Profits
GOP Test A Success

The *New Profit Times* reports that not-for-profits are getting mixed results on 900 numbers.

In Boston, the WGBH Education Fund raised over $50,000 recently by using a 900 interactive game in conjunction with the educational television station's annual auction. However, the Alaska Conservation Foundation just broke even when it used a 900 number in an attempt to raise money after Alaska's oil spill. The American Red Cross has had much greater success raising money for disaster relief using an 800 rather than a 900 number.

Larry Lynn, Channel 13's senior vice president of the Public Broadcasting Station in New York, reported at the *4th Media Journal* Conference that "we were extremely happy with our fund raising efforts utilizing 900." Optima Direct, who managed the NRPC campaign, did the Channel 13 campaign.

"The business lends itself so well to fundraising that it's a natural," according to Dan Weintraub, director of Sales and Marketing for Gateway Telecommunications, Inc. of Fort Washington, Penn. "We've served as a service bureau for a variety of successful fund raising campaigns. All the donor has to do is make the phone call and pay their phone bill," he said.

Some telephone companies are reluctant to bill for candidates, but the National Republican Party Committee (NRPC) did a successful July test with a 200,000 piece mailing, half of which included a $10-per-call AT&T 900 number. The actual returns are proprietary, but the NRPC does feel that when you combine a 900 direct mail piece, the response is higher. Kevin Potter, of Optima Direct of Washington D.C., stated, "the Republicans will continue to test and refine 900 as a political tool." *Oct. 1990*

900 Memorial Donations

Heather Carmichael and her husband were planning to do a Canadian financial pay-per-call line when at a funeral they were asked to make a donation in lieu of flowers. This gave them the idea for Care-Tel. The Carmichaels took a year putting the line in place and getting the charities on board. Now Carmichael said, "They are beating down my door trying to sign up."

Within the next two months they will join with a U.S. service bureau and AT&T to bring this service to the states. They and the charities are eagerly awaiting the start of nation-wide 900 in Canada and will switch their local Canadian Bell lines to 900. Joint media ventures are being planned.

This Canadian phone fundraising application allows callers to make a memorial donation on behalf of anyone to their choice of more than 33 charitable organizations. For example, an ad in the death notices section of *The Ottawa Citizen* advises readers to select any of three local 416 lines for donations of differing amounts --$25, $36 or

$50. Other donations, in different amounts are available in other ads. The five-digit access codes allow for tracking ads and for determining the amounts each of the charities receives.

The funds go into a trust account and Care-Tel receives only normal service bureau fees for its efforts. At this time there are no entry fees to participate in the program.

Here is a sampling of the organizations the Carmichaels signed: Arthritis Soc., Asthma So., B'nai B'rith Canada, Canadian Fund for AIDS Research, Canadian Breast Cancer, Canadian Cystic Fibrosis, Canadian Diabetes Assoc., Canadian Liver, Canadian Mental Health Assoc., Cerebral Palsy Ont., Hospice Ont., Multiple Sclerosis of Canada, Muscular Dystrophy Assoc. of Canada, Ontario Cancer, Parkinson Foundation of Canada, Schizophrenia of Canada, The Salvation Army, Jewish Family and Child Service, Mount Sinai Hospital, Sunnybrook Health Science. Centre, and Women's College Hospital. *Jan. 1994*

Dog Sled 900

Chalk up another 900 line with ties to historic expeditions. Norman Vaughan, 88 and chief dog musher from Admiral Richard Byrd's 1928 expedition to the South Pole, is making that journey again. Except that now in 1993, the public can join in by calling 900-990-534, ext. 210, to hear progress reports for $2 per minute.

Calls run no longer than five minutes and are updated twice weekly with the proceeds helping to fund the expedition. They are scheduled to reach the South Pole this month. *Dec. 1993*

Easter Seal & Vari-A-Bill

The Easter Seal Society of Utah is using AT&T's Vari-A-Bill service for fundraising efforts. The cost of a call to 900-884-SEAL is $50, but with Vari-A-Bill, the cost can be raised to $75 or $100, at the option of the caller. Vari-A-Bill (*AN* Dec. '92) offers five rate options on AT&T's MultiQuest® 900 service.

After the original five-day trial during the annual Easter Seals Telethon, it was decided to continue the effort throughout the year. "The overall concept of a pay-per-call system for charitable contributions was appealing," said David Miller, vice president of development and community relations for Easter Seal's Utah headquarters. "But having several different 900 numbers at different prices seemed too unwieldy. We wanted people to have a choice regarding different donation levels simply by calling one phone number."

Las Vegas-based B.F.D. Productions acts as the service bureau for the 900 line. *April 1993*

Three AIDS/HIV Not-for-Profits Go 900

The seriousness of AIDS has impacted on 900. At least three 900 applications are raising funds for HIV/AIDS patient advocacy groups.

The New York-based Long Island Association for AIDS Care (LIAAC) operates 900-820-3999. Callers can make $12 donations to the agency. In the Philadelphia region, We The People Living With AIDS/HIV of The Delaware Valley operates 900-370-PWAS, where each call costs $10.

AIDS Support Companionship Services (ASCS), based in South Florida, uses 900 as a voice-mail option for subscribers to its publication for people with AIDS or HIV, The Companion. The $1.99 per minute line, 900-WE-ENJOY, benefits ASCS and other AIDS-related organizations.

Advanced Telecom Services Inc. operates as the service bureau for all three lines. Gail Barouh, PH.D., Executive Director of LIAAC, noted why hers and other organizations are turning to 900. "As funding shrinks in the face of mounting caseloads, non-profit agencies will be forced to become more creative and aggressive in their funding strategies." *Dec. 1992*

AIDS Line Relays Vital Info

Similar to the AIDS-related applications reported last month (*AN* Dec. '92), the American AIDS Health and Information Program has a fundraising aspect.

The line (900-884-AIDS) costs $1.99 per minute, and IdealDial, the service bureau, reports that a substantial portion of the proceeds goes to benefit the Center for Disease Control's research for a preventive vaccine. The message, updated regularly, reflects new information available as AIDS research progresses *Jan. 1993*

FDESA 900 Number

On January 17 (Martin Luther King Day), a 900 number went online for callers to make a $15 contribution to the Fund for Democratic Elections in South Africa (FDESA).

Callers to 900-8-VICTORY hear a message from FDESA National Co-Chair Danny Glover and African National Congress President Nelson Mandela. FDESA is a non-profit, non tax-exempt corporation with the sole purpose of raising at least $1 million from contributions throughout the United States to support partisan free and fair elections in South Africa.

FDESA is asking for the help of the 900 industry, both in contributions and involvement. They are urgently looking for ways to publicize the number. If you can help in any way, call Wendy Swart, campaign manager, at 617-437-6363. *Feb. 1994*

Quebec Political Party Uses Fundraising Line

Canadians who wish to make a $50 donation to the Canada Party of Quebec can call a 900 number. Handouts by IP Tony Kondaks announce that the $50 membership enables Canadian citizens "to participate directly in the fight against separatists." *Sept. 1994*

High Seas 900

In what a famed sailor has termed "the highest and best use of 900," 900-820-BOAT provides daily progress reports of two Americans' attempt to break a 140-year-old sailing record.

In Ocean Challenge, famed sailor Rich Wilson and co-skipper Bill Biewenga are sailing to Boston, going around Cape Horn.

Calls cost 95 cents per minute. The service bureau is West Interactive Corp. Information for the line is gathered from the sailors' daily radio calls to relay stations. High seas operators patch the calls through to West Interactive, which records the updates and puts them on the 900 line.

"Our objective is to share the adventure taking place at sea with as many people on shore as possible," said Lyon Osborne, project manager for Ocean Challenge. "The only way Ocean Challenge can afford to make the daily reports available nationwide is through the use of 900."

A portion of the proceeds from the line is going to the Newspapers in Education program and the American Lung Association. The voyage it chronicles is expected to end in late March. *Feb. 1993*

Three Non-Profit Lines -- Whales, AIDS & TV

Three 900 lines from non-profit organizations, all managed by Lynn Adams Communications, Inc., former AT&T MultiQuest® manager, are being serviced by Sprint Telemedia. The lines originate from the American Cetacean Society (ACS), the American Social Health Association (ASHA) and the Television Viewers of America (TVA).

When asked why these lines were placed with Sprint Telemedia, Adams said, "The requirements for public service advertising and Sprint Telemedia's unique flexibility make them the only game in town." Lynn Adams Communications *(AN,* Feb. 1992) helps charities and non-profit organizations raise funds through 800 and 900 Telemedia programs.

The ACS line (900-535-9425) offers callers information on how to help the society's cause of protecting whales, dolphins, porpoises and their habitats all delivered by actor Leonard Nimoy. Calls cost $2.50 a minute and provide callers with the opportunity to leave their name and address and receive a complimentary mailing.

The ASHA Information Line at 900-535-2437 offers callers information on AIDS and other sexually transmitted diseases. Each call costs $10.

The TVA's "Petition for Competition" line at 900-230-1221 allows callers to learn about cable television industry issues and the position TVA advocates on consumers' behalf. Each call costs $2.40 per minute. Callers can leave their names and addresses for a Petition for Competition for Congressional Representatives. *Sept. 1992*

Chapter 14
News, Politics & Opinions

Opinion polling was one of the first 900-number applications, during the Reagan-Carter debates in 1980, and continues to be a popular method for getting instant readings of people's opinions on a wide range of topics.

Opinion polling need not be limited to simply casting a vote by calling the appropriate 900 number. *Newsweek* magazine has been using a 900 letter-to-the-editor service for those opinionated souls who are too lazy to sit down and compose a letter. Or to be a bit more charitable, perhaps they're simply poor letter writers.

Air & Space Newsline
Updates As Needed

Since the end of September, McGraw-Hill Aviation Week Group using Sprint Gateways has been operating a 900 line providing round-the-clock Aviation/Aerospace news for aviation and aerospace managers, news media and interested professionals.

Messages are changed daily, but Michael Miller, an editor at McGraw Hill, who had the idea for the number said, "At least every other day we use the preamble to present late breaking news throughout the day."

Each taped message includes summaries of each day's stories from Aviation Daily and Aerospace Daily, the daily intelligence newsletters covering the industry worldwide plus news as it happens from Washington and other news centers. Callers are given the option of hearing either of two news briefings: one for commercial aviation and the other for defense and space.

Mobile aerospace executives not at their desks have the option to hear the news highlights for two dollars a minute with an average call-time of three minutes.

McGraw Hill publicizes 900-233-2700 through its publications, using 1/4-1/2 page ads. "The number serves as a marketing device to sell the newsletters, allowing potential buyers to call the number and get a flavor of the newsletter," said Miller.

"Every comment about the line has been positive," and, according to Miller, "we are hopin.; to raise the call-volume sufficiently to drop the cost. Call-counts were not as high as projected for the first month," Miller believes, "because of the amount of 900 blocking in many of the businesses anticipated to use the line." *Nov. 1990*

Taking a Fling for the Fall of Fidel

Investment Vision reports that The Council for InterAmerican Security, a conservative group that believes Cuba is ripe for revolution, is holding a "Predict the Fall of Fidel" Contest.

Pick a date and dial 900-998-2800, extension 100 ($3 a call). The grand prize is a three-day vacation in Havana; second prize is a trip to D.C.; third is a box of Cuban cigars. *April 1991*

Guvs Veto 900-Governor

A new National dial-it service, called 900 Governor, received a good deal of free advertising, when Planned Parenthood took out ads in various papers requesting callers to express their support for pro-choice.

However, some Governor's argue that a phone call to the 900 number priced at $1.95 for the first minute and 95 cents for each additional minute, "was unnecessary and inefficient" as calls could be made directly to their offices.

Tennessee Director of Consumer's Affairs, Elizabeth Owen complains that callers are not told the price of the call prior to leaving their message.

While Creative Marketing Associates president Maynard Small asserts that "this service is perfect for the start of the 90s." Ms. Owen, contends that this service is "geared to Mr. Small's pocket book."
April 1990

CNN's Interactive Program

By the end of July a 900 program will enable viewers to vote for the news stories they want to see. In a first for television news, CNN has developed the Newsnight Program, which will open with the top six stories of the day in 25-30-second summaries. For 95 cents the callers vote for the story they want covered that very evening.

This appears to be the logical outgrowth of the 900 telephone polls that have given viewers opportunities to interact with television. *July 1990*

Viewers To Choose Stories At CNN

The first experiment in interactive news by a national network, Cable News Network, has started. At the beginning of the show, viewers see a summary of headlines and as anchors read the day's top news, viewers can dial a 900 telephone number for 95 cents to vote on which of the peripheral stories they'd like to see. Although viewers may feel they have a decision-making posture in what gets aired, CNN insists that major stories (as determined by the news executives) will run as they always do. *August 1990*

News From Home Travels Via Fax & 900

Hartford Courant readers away from home can get their local news fix via 900 by punching in the nearest fax machine number and they will immediately receive the daily FAXPaper. Instant Information, a Boston based 900 information services company has launched the U.S.'s first 900 FAX-It-Back newspaper.

The FAXPaper is a capsuled one-page news summary directed at businesses, banks, insurance companies, small businesses and public officials within the Hartford, Connecticut area. The daily is available immediately from any location, at any time. *Oct. 1990*

New Republic Goes 900

The influential public affairs journal, *New Republic*, has seen the light and gone 900 -- welcome. The $1.25 number takes letters and other views. Proceeds will go for internships and other special editorial projects. *April 1991*

Newsweek Turns to 900
Uses MCI's Letterline for Letters Column

Believing that for some readers, putting a pen to paper is too slow a vehicle to get their message across, *Newsweek* magazine introduced a telephone mailbox. The charge for the service is $1.95 for the first minute and $.95 for each additional minute.

Diana Pearson at *Newsweek* said they are not replacing the written letters but are using the 900 number as a convenient way of drawing in additional reader comments. When *AN* spoke to *Newsweek*, the line had received a few hundred calls in less than two weeks.

Some of the messages are incorporated into the LETTERS column, identified by the symbol %. *Newsweek* is the first to use LETTERLINE, MCI Information Resources' new service. The MCI subsidiary which developed the service maintains the program software, hardware and call answering facility. The company also produces and updates the opening greeting; recruits and trains message screening personnel; monitors and sorts messages; allows client access to the system and provides copies of comments on tape or transcript.

Companies sponsoring a LETTERLINE set their own charges. *June 1991 (active)*

Press Releases Sent Via 900 Fax
Miami Herald Uses Fax with 900 Number

On Monday, July 8, when you tried to fax material to the *Miami Herald* using the wire room's fax number, there was a request to pick up the handset. The message: starting that day, unsolicited press releases to the *Miami Herald* would be pay-per-call.

MCI had converted the *Herald's* fax number to 900 (900-988-4329), and for $2.00 the first minute, and a dollar for every subsequent minute, callers could fax away.

The Herald decided to do this to save time, and fax paper, and to escape from the blizzard of unsolicited press releases.

On day one they received 642 phone calls and Peter Schaeffer, the wire room supervisor, admitted that this might well turn out to be a money maker for the paper. *Aug. 1991 (active)*

Interactive Public TV

The 1.7 million viewers of WLIW Public TV, in Long Island, NY, can vote for their preferred program by dialing 900-288-WLIW. Each month, callers are given a new menu of programming options, and the program that gets the most votes is aired on a specific evening that month. The cost of each call is $2.00.

Jessica Hillman, representing the station, which is among the ten most watched public TV stations in the country, said, "Viewers Voice is going very well."

The station is using IdealDial, a Colo.-based service bureau. When *AN* asked how they had made their choice, Hillman said, "We met them at a public broadcasting conference in Boston." *March 1992*

Miss People's Choice Honors Via 900

In a pageant created by GuyRex Organization, known for shepherding a string of beauty queens to national titles, the winner of the people's choice honors was selected by 900. Miss El Paso won the award by earning the most votes from viewers calling in the 900 number. *Sept. 1991*

Loan Guarantee 900 Number

The America-Israel Phone Project has set up a 900 number, that for an $8.98 fee, will send letters to the President and members of Congress urging prompt approval of $10 billion in loan guarantees for Israel.

The number, 900-PRO-JOBS, features Monty Hall, who describes the economic distress of hundreds of thousands of immigrants who came to Israel from the former Soviet Union. *April 1992*

Tax Protests Make Money
Audiotex Mobilizes Political Opinions

Coventry Communications of Fort Washington, Penn. has developed the software that enables it to conduct statewide public opinion polls, which it calls advocacy campaigns, using a 900 number. Coventry's application gave voters of New Jersey an opportunity to send positive or negative letters to three local legislators and the Governor.

They were hesitant to release the numbers, but they are going to continue in this niche market. Coventry's software goes beyond New Jersey giving them the ability to identify local legislators from an address in any of the states.

New Jersey is in the midst of a taxpayer revolt, with a Governor who is perhaps the most unpopular one in recent history. To advertise the application, Coventry utilized a press conference timed to take advantage of a speech the Governor was making to explain the higher taxes. They made virtually all the papers. *Oct. 1990*

800 to 900 Encourages Pay-Per-Use

It cost the state of North Dakota $75,000 for an 800 number for the 1989 legislative session. North Dakotans can now contact their legislators and express their opinions for a 25 cent 900 call. *Nov. 1990*

Opinions Count on 900
Helps Develop Data

Using a simple dial and punch, with one extension for yes and another for no, people can make their opinions count. Widely available for political campaigns, the system has been used to compile data for other issues. For example: in Long Beach, Calif., to register opinions on whether Disney should build a theme park, in Westchester County, N.Y., for views on the construction of an airport, and in New Jersey on whether to recall the Governor.

IPs can assist not-for-profits in getting the hard data they may need to present a community viewpoint. This use of 900 helps the public see useful applications, provides consumer awareness about how the technology operates, and helps present a positive industry image. *May 1991*

Sound Off on Negative Ads

The League of Women Voters started a telephone complaint line during the New Hampshire primary. Callers to 900-Run-Fair can report campaigns ads they think are overly negative or unfair. The call costs $1.15 per minute and is cosponsored by the American Association of Advertising Agencies with funds from the Markle Foundation. *April 1992*

Candidates Try Audiotex
Ground Rules Developing

Former Atlanta Mayor Andrew Young used a 900 number and a Voice Box number for his campaign in the Democratic Primary for Governor of Georgia. Young received a $10 donation for each call to the 900 number. His manager, Hobby Stripling, believed that Young's prestige and his "star quality would override the 'talk dirty' image that some of the 900 numbers promote." *Sept. 1990*

Meet The "Real Politician"

Callers to 900-820-CNIP ($3 per call), learn how national groups view politicians and how much money political action committees representing agriculture, business, labor, etc. gave lawmakers. *Nov. 1990*

Tel-A-Letter
Audiotex Gives Voice

Tele-Serve, San Francisco service bureau using AT&T lines, is helping a group of entrepreneurs in Idaho develop a Tel-A-Letter 900 application.

This application provides a way for individuals to write politicians and others in power positions. Concerned individuals can call one of several Tel-A-Letter "Action Numbers," each corresponding to a specific issue -- abortion, waste recycling-- and respond to a series of questions which will generally determine the callers' personal feelings on that issue. Computers assimilate the answers to these questions and four one page letters are drafted to four different predetermined decision makers. For a cost under $8, a personalized letterhead is created and the letters printed. They are sent to the caller for his/her signature, P.S., and mailing to the individual addressees as shown in the letters. *Sept. 1991*

News From The Holy Land, for $1 A Minute

Dateline Jerusalem Inc. of Palo Alto, Calif., is offering a specialized 24-hour news service through 900-568-News. Callers hear recordings of broadcasts of The Voice of Israel and Israeli Army Radio in either English or Hebrew, for $1 per minute.

Harry Saal, of Network General, a high technology firm in California and Bruno Wassetheil, a former CBS News radio correspondent assigned to Israel from 1970-1986, are heading up this effort. *June 1990*

Israeli News Live Via 900

As previously reported in *Audiotex News*, Dateline Jerusalem Inc., of Palo Alto, Calif., offers a specialized 24-hour news service through 900-568-News.

Callers hear recordings of broadcasts of The Voice of Israel and Israeli Army Radio in either English or Hebrew, for $1 per minute.

AN found that during the Persian Gulf War calls jumped from 2,000 a week to 2,000 per day. Callers were pleased to have a real sense of "being there" and the different perspective the Israeli news teams provided. *April 1991*

Operation Desert Shield
New 800 & 900 #s Help Troops

With the U.S. troops stationed in Saudi Arabia in what President Bush calls Operation Desert Shield, 800 and 900 applications have been developed to help Americans keep in touch with the troops, access up-to-the-minute news and tell off Saddam Hussein.

Pennsylvania residents dialing 900-386-6800 can leave a non-emergency message for friends and relatives in the Middle East. Systel Communications Inc. of Yardley, Pennsylvania, who developed the program, transcribes the messages to an "Operation Reach-Out" telegram.

TelAd, of Lakewood, Ohio, has a 900 number designed to allow the public to tell Saddam Hussein what's on their minds. By dialing 900-468-8000, interested callers can get access to a number for expressing individual opinions. Each call costs $1.95 for the first minute and 95 cents for each additional minute. The maximum time allowance is five minutes.

The public can call Sprint Gateways' toll free number, 800-676-2255, for updates on the Middle East crisis. Callers will be faxed a news bulletin entitled "Crisis Update: The Persian Gulf" from FAXIS, a New York-based fax information service company.

Tapes from American radio stations, along with recorded messages from relatives and others back home, can reach the troops. A 900 number, 900-820-2USO, is available for callers who wish to record messages for the "USO Morning Show Network."

Military reserve employees of NYNEX involved in the call-up are supported by the company. Those involved will receive insurance plus normal salaries and military pay for six months. According to NYNEX chairman, William C. Ferguson, "No employee in the NYNEX family of companies need worry about either their job or the financial impact if they are called to active duty during our country's time of trial."

NYNEX has established an 800 number, 800-228-1524, for employee reservists information. *Oct. 1990*

Embassy Presents Political & Economic Info
Jamaica Uses 900 # Update

The Government of Jamaica is using a 900 telephone number for Jamaicans living in the US, as well as others, to keep up-to-date on the latest political and economic news from home.

USP, a Boston service bureau, assisted the Jamaican Embassy in Washington to set up the program on Sprint Telemedia lines.

The line began in September and is still in the testing phase. The number was marketed through press releases and public service announcements. The application has menus and sub-menus.

Wayne McCook, the Jamaican information attaché said the embassy is pleased. The number is 900-535-2200, and 95 cents per minute covers the government's costs.

Callers can leave their names and addresses on a mailing list for continued updates and action alerts. The number offers three sections: news and information for Jamaicans living in the US; investment and business opportunities in Jamaica; travel and visa information for those interested in visiting Jamaica. *Feb. 1992*

Amnesty ActionLine

Amnesty International's (AI) ActionLine has been up for more than a year now and is still going strong. Callers to 900-446-4020 can get AI's latest appeal for action faxed to them for $3.

The human rights group conducts a letter-writing campaign in its worldwide efforts to free prisoners of conscience and to ensure fair trials for political prisoners. Each faxed appeal contains news of human rights violations and directs AI members to send written pleas for cessation to officials in the country involved. It gives addresses and suggests points to stress in letters.

Gene Rosov, president of Innovative Telecom Corp., which operates the ActionLine system, said that the line allows AI members to take quick action in their letter-writing efforts. It serves another important function.

"We're talking about saving lives for $3," Rosov said. "That's a very good cost-benefit ratio." *Sept. 1992*

Canadian Candidate Uses 900

Kim Campbell is a candidate for prime minister in Canada. Her campaign seems to combine President Bill Clinton's politics of inclusion and presidential candidate Ross Perot's pronouncement that he wanted to give the country back to the people.

Campbell has established a 900 number and said, "... anyone in this country [Canada] can call it and find out how they can be part of what I hope is going to be an extraordinary movement to involve

Canadians. I have sensed the excitement across the country; and frankly, our party is the only party that is extending that invitation."

The number is 900-5615-KIM, and the call costs 50 cents for the first minute and 35 cents for each additional minute. *May 1993*

Dennis Miller Fans Call 900

Opinion polling via 900 marches onward. Last summer, as the Dennis Miller Show faced its cancellation, viewers were invited to call either of two Sprint Telemedia 900 numbers to say "if you think the show should stay on the air" or "if you don't think the show should be cancelled."

The cost of each call was 75 cents, and net proceeds went to benefit the Pediatric Aids Foundation. Due to the similarity of the questions, one could guess that this was all in fun and fundraising. Thousands of viewers felt strongly enough about the "issue" to call. *Oct. 1992*

Celebrity? Lines

Last month two 900 lines made news. One featured serial killer John Wayne Gacy, awaiting execution this month, offering his thoughts at $1.99 a minute. Callers can hear a 12-minute recorded interview in which Gacy calls himself the "the 34th victim" and claims someone else committed the murders.

Gacy and All Star Communication of Boca Raton, Fla. set up the line, which began operating March 1. Average cost is $23.88.

The other is the Derrick Smith and Shane Stant opinion line about the Nancy Kerrigan/Tonya Harding "saga." Calls to 900-388-1100 cost $4.99 a minute with the average length five minutes. The line, which began in March, is being used to raise money for legal bills.

The press printed the Kerigan/Harding number, but not the Gacy number. *May 1994*

Dateline Jerusalem Again
Per Minute Charge Goes Up

After ceasing operation last year, Dateline Jerusalem is again available at 900-737-NEWS, with recorded news items in English and Hebrew gathered from Kol Yisreal, the radio Voice of Israel. The service began in 1990 *(AN*, June '90).

News items are updated at least three times a day, and each call costs $1.50 per minute, with the average call running three to five minutes. When the service began in 1990, the cost was $1 per minute. Dateline Jerusalem is headed by Bruno Wassertheil, former CBS News Israel correspondent. According to Wassertheil, the line provides details and background information on news stories from Israel that ordinary news sources frequently omit or miss. *July 1992*

Falwell's Whitewater 900

Televangelist Jerry Falwell added a 900 phone line to his campaign to uncover facts about President Bill Clinton and the Whitewater controversy. The $1.95-per-minute line has been advertised on 225 television stations nationwide offering callers "shocking" information on Whitewater.

The line's taped message reportedly doesn't give much information, but allows callers to add their names and addresses to a petition calling for Congressional hearings on Whitewater. It also advertises Falwell's $40 videotape.

According to published reports, Falwell doesn't expect any profit from the 900 line or videotape sales, just funds to offset spending on his Whitewater campaign. *June 1994*

Fox's 900 Naming Polls

The Fox Broadcasting Company selected First Data Corporation's Call Interactive to process calls for a 900 number poll held late last year. The poll asked viewers to choose between two names for the new baby on the series *ROC*.

All month and during the birth episode, viewers were asked to call 900-454-ROC1 to name the baby Marcus or 900-454-ROC2 for Shaka. Each call cost 79 cents.

The poll generated a clear winner with Marcus receiving 21,749 votes. The other name, Shaka, received 8,692 votes. All proceeds from the calls were donated to National SAFEKIDS.

It seemed to *AN* the call counts were disappointing considering the aggressive advertising and promotional campaign used by Fox for the 900 poll. Advertisements appeared in trade and consumer magazines, including *TV Guide*. The network also ran promos featuring *ROC* star Charles S. Dutton.

In sharp contrast, in another Fox 900 application where early last month viewers decided which of three *Living Single* characters would go on a date with a new neighbor, almost 250,000 viewers called and voted. Proceeds from the contest go to Literacy Volunteers. *March 1994*

Free Press Plus 900

The *Detroit Free Press* took advantage of the presidential election to add to its Free Press Plus 900 applications. The Michigan paper offered a three-page fax comparing the candidates on major issues, ordered through 900-740-PLUS, menu option 7, for $2.50 per minute.

Also, a presidential debate forum line ran on the same 900 number, menu option 9. Callers could listen to others' comments on the campaign's debates and add theirs for $2.50 per minute. The paper reported that almost 100 callers dialed in after the first debate. The other Free Press Plus 900 applications cover weather, lottery, crossword clues, soap operas, stock quotes, car appraisals, sports and mortgage rates. *Dec. 1992*

Letters Sent
Public Speaks to President Via 900

The American public has lots of opinions to voice to the President, especially in an election year. Now, As *Stars and Stripes* reports, messages from ordinary citizens can get through on 900 lines. The X-Press Letters to the President phone line, allows callers to leave a brief message for President Bush on any subject they choose at 900-370-1000.

Gary Williams, an Illinois resident, began the $2.49 per minute line last July. According to Williams, who acknowledges being a Democrat, the line is non-partisan and intended to be unbiased.

The average call lasts two minutes. When transcribed from the recorded calls, the finished letters are forwarded to the White House. If they wish to receive a reply from the President, callers can leave their names and addresses. Threatening calls are turned over to the Secret Service.

Williams said that he expects 300 to 1,000 calls a day once the line really gets going. About 40,000 to 50,000 letters arrive at the White House per week. *Oct. 1992*

900 Poll on Homelessness

Similar to its 900 poll on aging *(AN* Jan. '94), *Parade* magazine conducted a 900 poll on homelessness in conjunction with a story on its homelessness survey.

Each call costs 75 cents. The poll asked callers' ages and for yes or no answers to questions about whether they would pay additional taxes to help the homeless, if there should be a law to remove homeless people from the streets, whether they worry about ever being homeless, and if they think most homeless are responsible for their situation. *Feb. 1994*

Line Carries News from Indian Subcontinent

Joining Dateline Jerusalem *(AN* July '92) with news from Israel, and news and sports' bulletins from Ireland *(AN* March '93), is Global News Service offering news features from the Indian subcontinent and Afghanistan.

Updated daily, the 95 cents per minute call to 900-737-0733 provides information not readily available from the mainstream American media. The service bureau is IdealDial. *April 1993*

Lewis & Clark Line
Touch of History & 900

The 900 industry recently gained an application with a connection to history. The Lewis and Clark Line at 900-I-GO-WEST gave daily updates of this summer's two-man expedition retracing the 1804 trek of 4,000 miles. Ten percent of the proceeds from the $1.95 per minute 900 line went to American Rivers, the non-profit conservation group that sponsored the expedition.

The Lewis and Clark Expedition 1992 ran from June 1 to August 13. Dr. Tom Warren and John Hilton followed the historic route. The purpose of the expedition was to investigate the changes that modern times have brought to the environment along the rivers and to educate the public about the need to protect our rivers.

Randy Showstack, American Rivers' communications director, told Audiotex News, "... the 900 number was a great way for the Lewis and Clark Expedition to keep the public updated on a daily basis on their adventures and the environmental problems the participants encountered as they went across the country." He added that the expedition was widely covered by the media, and callers to the line included the press seeking information to include in their news stories. *Sept. 1992*

Line Helps Americans Reach Legislators

Americans who want to voice their opinions to elected officials can do so through a 900 line established by the Alameda, Calif. company Public Correspondence. Callers to 900-776-0501 can dictate letters to the legislator of their choice. For a $15 charge, the company transcribes and mails the letters.

The line intends to get the democratic process moving, according to Jill Wilhelm of Public Correspondence, not promote causes. Since the line began earlier this year, letters have been on diverse topics.

Letters are in the callers' own words, and Public Correspondence will research a public official's name and address if necessary. *Dec. 1992*

Madonna - Yes or No?

With all the hype surrounding her, some of the public is sick of Madonna. The New York Post let readers express their feelings using a 900 number telephone poll. For 90 cents, sick readers could dial 900-990-9120, or Madonna supporters could dial 900-990-9121. *Dec. 1992.*

Mag Survey

Last month, *Parade Magazine*, in conjunction with an article on its survey on aging, advertised a 900 number for further polling. For 75 cents per minute, callers to 900-773-1200 were asked to use the keypad of their touchtone phone to answer five questions at 75 cents a call. The questions ranged from "How old is old?" to asking the caller's own age. Calls were taken between 8 a.m. Eastern Standard Time on Dec. 11 and midnight on Dec. 15. *Jan. 1994*

Messages Sent Via 900 to Officials

A September 1991 *AN* article prophetically noted, "Working out of a Ketchum, Idaho office, they [Tel-A-Letter] are a professional, talented group with interesting backgrounds. This idea has been tried before but not with the approach of marketing to special interest groups and IPs. The sophisticated software, the professional packaging and marketing presentation may make this a winner."

Last month, when *AN* spoke to Jeff Miller, president of Tel-A-Letter, it appears the prediction will be realized. With two years of experience behind them, the group is about to launch an "elephant size promotion," that Miller, understandably extremely excited and optimistic, predicted will generate a couple of million calls within six months.

While other earlier 900 lines, such as X-Press Letters-900-370-1000, allow callers to send messages to elected officials *(AN*, Oct. '92), Tel-A-Letter has a sophisticated approach that provides personal forms of communication.

Another group, U.S.A. Letters, is helping The Fleet Reserve Association encourage retired and active duty Navy, Marine and Coast Guard personnel to contact their elected officials and voice opposition to a presidential proposal and U.S. House and Senate Budget Resolution concerning military pay and benefits.

Last month, the association's Legislative Hotline began operation at 900-896-1776. For a $4.95 fee, callers leave their names and addresses, and within five days receive letters, ready for their signatures, addressed to their senators, representatives and President Bill Clinton.

At the time *AN* spoke to the association, no performance results on the line were yet available. However, Mike Zabko, director of membership development, said that a direct mail campaign had stirred up interest. Zabko noted that the association has a mailing list of contacts in military command centers, and the direct mail campaign encouraged them to call the hotline and ask their troops to call as well.

The association is also running ads in two magazines -- *Naval Affairs Magazine*, and *On Watch. May 1993*

More Opinion Polls

The use of 900 for polling purposes appears to appeal to the IP. Some examples:

In July of 1993, W&H Enterprises of Vero Beach, Florida did a two-week opinion poll on a Disney project in the area. Calls cost $1.45.

Also in July, Richard Bukowsky did a telephone poll dealing with the controversy over a school superintendent for $2.95 a phone call.

In January, Maramy Company, based in Crete Illinois, did a poll on the building of an airport in the suburbs.

Of course IPs who can manage joint ventures with newspapers have the best chance of getting good call counts over time. Even if the newspaper has a small circulation, the low advertising and low line maintenance costs can yield not spectacular but steady profits.

The *Daily News Journal* in Murfreesboro, Tennessee now allows callers to dial 900-289-1225 to give their opinions on local issues on a line co-sponsored by new local telecommunications company COMCORD-900. The newspaper has a circulation of approximately 15,168.

When the line began last March, its first question sought opinions on a proposed city and county school systems consolidation. COMCORD-900 grew out of a local group called CORD (Counsel on Redemptive Dialog). The Daily News Journal's 900 line is intended to help readers get a feel for what's on other readers' minds, said COMCORD-900 founder Tom Christy. *May 1994*

News From Erin
News & Sports Items

Similar to Dateline Jerusalem *(AN* July '92) with news from Israel, two 900 lines have started to bring news and sports bulletins from Ireland. Just in time for St. Patrick's Day, the 99 cents per minute lines at 900-420-2411/2412 carry news and sports items, updated several times a day, from Dublin.

Calls run about three minutes. The IP, Ireland's national broadcasting service, RTE (01 64 2350), uses Florida-based Interactive Communications Network as its service bureau. *Mar. 1993 (active)*

Polling Still Popular

Opinion gathering and "seizing the moment" continues to be part of the 900 industry, as shown by several recent lines.

The *St. Petersburg Times* let its readers get in on the Jay Leno/David Letterman debate. Through five 900 numbers, readers reported whether they preferred Letterman or Leno, thought that Letterman should stay on NBC and have his show pushed up to Leno's time slot, favored Leno to be aired at 10 p.m. and Letterman at 11:30 or vice versa, or preferred Johnny Carson.

The majority of viewers of ABC's *Home Show* thought Howard Stern should be taken off the air, according to a recent 900 poll.

Another line recently let callers say good-bye to former President George Bush. Before Bush left office, Dial-the-Prez, at 900-370-1000, took messages for Bush or Vice President Dan Quayle. The call cost $2.49 per minute, and messages were transcribed and sent to the White House.

A Seattle IP launched "Citizens Poll," using as his first question, "Should we allow gays in the military?" Each call cost $2.75, and phoners were promised President Bill Clinton would receive the results.

Polls appear to be an institutionalized use of 900, particularly when we recount as the 900 beginnings -- the Carter-Reagan debates in 1980, and the 1982 *Saturday Night Live* poll to boil or save Larry the Lobster. *Mar. 1993*

Write-It for Messages to Elected Officials

Similar to X-Press Letters is WRITE-IT, 900-737-3007, where callers have the chance to leave messages for the elected official of their choice. Limited to two minutes, each call costs $7.95. The caller's message of up to 40 words is transcribed and sent to the desired official.

The non-partisan line was started by two Houston residents, David Pena and Harry Rogers. *Oct. 1992*

Polling Lines Competition

Polling lines get competition from newspapers like *THE ENQUIRER* when they charge 39 cents. That's what the paper charged for each call in a recent phone-in vote for the Academy Awards best picture, best actor, best actress poll. *April 1994*

Speeches by Fax

A 900 number allowed callers to receive a fax of the full text of the speeches Barbara Bush and former president Reagan made at the Republican convention. Each call to 900-737-0700 cost $3. *Oct. 1992*

Celebrity Talk Show Host on 900

Talk show host Vicki Lawrence joins the celebrities having 900 numbers. Her 900-28-VICKI allows callers to leave a comment or story idea for 75 cents. The preamble features a fast-talking Lawrence and asks callers to leave their names and phone numbers before voicing their ideas or comments. *March 1994*

ACLU Offers Court Analysis

The American Civil Liberties Union just introduced 1-900-288-ACLU, at $1 the first minute and 75 cents each additional minute, so you can listen to its analysis of Supreme Court decisions or receive updates on relevant issues before Congress.

Former Billionaire Boys Club Head on 900

A 900 number allows Joe Hunt, convicted murderer and former head of the Billionaire Boys Club, and wife Tammy to get their story to the public. The $3 per minute line at 900-TALK-BBC was reportedly initiated through the joint efforts of an expert witness for the defense and a former juror nonprofit Billionaire Boys Club, and wife Tammy to get their story to the public. *October 1994*

O.J. Lines

The O.J. Simpson murder pre-trial hearings generated a number of O.J. 900 polling lines, joke lines, etc. The distribution of flyers advertising 900 and 809 numbers reportedly occurred outside the Los Angeles courthouse holding the proceedings.

Van Nuys, California-based JJ Cole Productions had excellent press for its O.J. line, including an article in the *Odessa American* in Odessa, Texas where IP Brian Cole resides. The line's number, 900-835-1414, ext. 114, was mentioned in the four-column article. Callers are asked to leave prayers and words of encouragement for Simpson. Cole intends to edit messages received through the $2.49 cents per minute line and forward them to Simpson.

Despite the press coverage, when Cole spoke to *AN* last month, he still hadn't gotten any calls. Cole said "I am reluctant to advertise at this time." Cole's company owns two other lines, purchased through Los Angeles-based American Information Marketing (AIM). The other two are a line for artists' copyright information and a political opinion line. *Sept. 1994*

Chapter 15
Miscellaneous

Finally, the catch-all chapter for those applications that simply wouldn't fit anywhere else in this book. Or those applications that are too broad to be slotted into only one category.

Caller Billed For Cellular

For some time now there have been attempts to have the calling party pay for calls made to the owner of the cellular phone. Don Young, of National Phone Programs, believes there have been 5 different moves to have the callers billed, none of which have really worked out.

However in Houston, there is a promising proposal that may turn out to be a lucrative application for 900. VYVX Telecom Inc. an Inter-exchange carrier worked with Houston Cellular to develop the product known as Calling Party Pays. Calling Party Pays is a service provided by VYVX Telecom Inc. to Houston Cellular customers.

Subscribers to the program, which started the end of April, pay a $10 monthly access fee, in addition to normal access fees, and are reassigned a cellular 900 phone number. Intrastate calls made to that cellular phone number while the customer is in the Houston coverage area are charged back to the caller at a flat, use-per-minute charge. Later this year, the program will be expanded to include calls from anywhere in the continental U.S. *July 1990*

Bees Buzz Information

The *Sacramento Bee's* entry into the field of audiotex began with a wide range of news, reviews, sports and entertainment. An interactive service, the caller after dialing 552-5252, at no charge, can select from the directory a variety of topics including state and local news, Giant's and A's scores, local weather, lottery results, movie reviews, quiz games and realtime stock prices. *Audiotex News* will watch the application to see if it will go 900. *June 1990*

NY Times Loves 900
All The News Fit to Dial

Audiotex News believes *The New York Times* has embraced 900, not only in a big, but in a creative way. Although it is fairly secretive about what's going on, we do know it has been developing applications one after another. That seems to indicate that existing ones are paying for themselves, but all Nancy Neilsen, *The New York Times'* head of corporate relations, would say is, "900 is a way of improving service to readers."

It began with what in the future will be called a crossroad in the industry, calling for clues to its famous crossword puzzle. Starting last November, if you were so inclined, you could get the Bestseller List of books 10 days before publication. Subscribers automatically receive a one-page fax of the lists every Thursday morning at 8 o'clock for $325 per year. For fax on demand, you dial 900-773-Fax at a cost of $ 7.50 per page. If you call 900-454-List, for 1.50 per minute, you hear the Bestseller List.

Restaurant reviews are 75 cents a minute (900-988-0101), and can be selected by neighborhood, type of cuisine and price level. According to Neilsen, they will be expanding this to cover more than the current 250 restaurants. Callers get data on atmosphere, recommended dishes, credit cards, hours and reservation procedures.

The Times' most recent addition will be competition for the American Express USA-Weather line. The Times number is 900-884-Cast, and for 75 cents a minute, you can get a report that is updated hourly, but if conditions make it necessary, every 15 minutes. On the same number you can also hear ski conditions for 160 ski resorts, which turn to beach and boating reports in the summer. "Holy

Cow" what's next -- *Audiotex News* readers will be the first to know. *Feb. 1991*

ANI Begets Two #s

The Private Lines Company, of Beverly Hills, California has developed two applications, 900-Stopper and 900-Runwell, that allow users to block the transmission of their own telephone numbers in calls anywhere in the world. Automatic Number Identification (ANI) subscribers see only a string of zeros on their display screen.

Both of these numbers use the same telephone switching computer in the Midwest. A caller to either 900 number receives a second dial tone and then dials the number wanted for $2 a minute for the national and $5 a minute for the international. AT&T provides the inbound and outbound long-distance telephone service. *May 1990 (900-Stopper line active)*

Nuclear Physicist Guides Callers

Entrepreneur Ryan Wood, of Menlo Park, Calif., president of The UFO Line, is so convinced that there is an unserved market for reporting and hearing reports about UFOs that he has Stanton T. Friedman, a nuclear physicist and nationally known UFO expert, guide callers through a menu of extraterrestrial topics.

Options on the keypad include: recent sightings, physical evidence of UFO reality, flying saucer technology, and information on "UFO abductions." The call-cost for 900-USA-UFOS is $2 for the first minute and $1 for each additional minute. *July 1990*

900 Used To Connect Thousands

Through a 900 broadcast, distributors of Herbalife, a weight-control powder, met by phone. More than two thousand callers tuned in. The 30-minute call cost distributors an average of $13 -- representing a saving in time and money over attending the meeting in person.

Using AT&T's MultiQuest® Broadcaster, 900 numbers are used for domestic customers, salespeople and distributors. Special calling periods are reserved for Spanish, Korean, or Japanese-speaking callers. *Sept 1990*

Phonecaster
Thousands Now Can Hear Live Events

Callers can now listen to live events such as sporting matches, debates, conferences, concerts, shareholders' meetings, space shuttle launches, or any event, through a service offered by Scherers Communications, Inc. called "Live Phonecaster."

This is a one-way live broadcast which employs an inbound 800 number for the live feed of the event. Callers, who may dial in at anytime during the event, use an 800 or 900 number. The cost of the call can exceed one dollar a minute.

The cost to prospective IPs includes a set up fee of $500 for a one-time event and $100 for a recurring event. Usage fees for inbound 800 are $.40 /minute and outbound 900 are $.15/minute. *Sept. 1990*

Nationwide Quilt Number

900-USA-QUIL lists nationwide quilt shows, workshops, fabric sales and guild meetings. For $2.00 the first minute and $1.00 each additional, Sara King's recorded message also provides a quilting tip. King is donating 10 cents for each minute to Holt International Children's Service (an agency for homeless children overseas). *June 1991*

Playboy Wake-Mate

Playboy Wake-Mate, recently announced by Playboy Enterprises Inc., offers a one-minute recorded wake-up call from a *Playboy* centerfold for $4.

Callers dial 900-820-WAKE, enter their phone number and the time they wish to be awakened.

The morning messages include such goodies as: Good Morning, sleepyhead. Did you have a good rest? Come on up and at 'em. You'll knock them dead today. *March 1991*

Teleline's INN-TOUCH
Tour & Travel Info Services Network

In October, Teleline Inc., a California Service Bureau, introduced their INN-TOUCH service at a convention of Best Western operators.

This marks the debut of this information service which accesses up-to-the-minute local and national weather reports, sports information, financial data and local entertainment and event information.

Guests are billed 95 cents per minute and can reach the service by pressing an "I" and "T" on their room phone. *Nov. 1990*

Guitar "A" Noteworthy Program

Through one of the nation's few audiotex programs that do not use the spoken word, guitar aficionados can now "Dial An A." Designed to assist in guitar tuning, 900-903-TUNE elicits an "A" tone from which guitarists tune their instruments.

Calls to the program, introduced by Integrated Digital Systems in conjunction with Advanced Telecom Services, cost 99 cents per minute. Each call has a two-minute maximum.

Initial results are good, and expansion to include the tones for the B, C, D, E, F and G notes is being considered. *July 1992*

Newsday and NYNEX Team
Newsdial Starts

Newsday, a Long Island, N.Y. newspaper that has consistently bathed the 900 industry in a negative light, set up an electronic publishing department and last month introduced Newsdial-- a service that provides everything from sports news to career advice and gardening tips over the phone -- using 900 numbers.

In a noteworthy alliance between the publishing and telephone industries, *Newsday* joined with NYNEX Corp. to begin a six-month test of a system for providing news over the phone this fall. Subscribers to New York Telephone's Call Answering service would be able to receive news briefs from *Newsday*.

The purpose of the test is to determine how much users are willing to pay for such a service, what type of news to broadcast, and how to provide access to the service.

Newspapers 900

More and more IPs have to compete with newspapers who are providing 900 pay-per-call services.

The Cleveland Plain Dealer (daily circulation, 400,000, Sunday 500,000) offers sports news and scores (900-370-3300) *(active)* updated every 15 minutes for 99 cents a minute. Other lines include Jeanne Dixon horoscopes (900-988-7788) *(active)*; the Horse Race Results Line (900-976-0789) *(active)*, a service of Conor Communications Co., with race results from tracks nationwide; Lottery Line (900-860-3350) *(active)*, offering numbers from 31 states; the Accu-Weather Hotline (900-990-8506), with forecasts for more than 2,000 cities worldwide; Crossword Clue Line (900-896-2123) *(active)* and Soap Opera Update Line (900-329-7676) *(active)*, all for 95 cents per minute.

The paper also offers Voice Greeting Card (900-726-1121), where callers can send a greeting for any occasion for $1.25 per minute, and Person to Personals (900-737-3307) for $1.79 per minute.

According to a report in Information Industry Bulletin, the largest ancillary service at the New York newspaper *Newsday* (daily circulation, 763,000, Sunday, 852,000) is audiotex. The paper has a policy of only expanding into ventures that will be profitable within one year, and president and publisher Robert Johnson reports that the figures for audiotex volume have been running ten times ahead of expectation.

Newsday's 900 lines include: Newsday Highlights (900-BULLETIN) offering highlights from the paper for $2 per minute; Weathertrak (900-820-1900) *(active)* for domestic and foreign weather reports at 75 cents per minute; Sports Now (900-680-1234) *(active)*, scores and statistics for 99 cents per minute; nationwide horse racing results, (900-933-4343) *(active)* for 95 cents per minute; Lottery Results (900-622-6886) *(active)*, New York State Numbers for 60 cents per minute; Marketline, mutual fund and stock prices (900-448-2442) *(active)* for 75 cents per minute; Soap Opera Update (900-329-4558) *(active)* for 95 cents per minute; Crossword Help (900-285-6456) *(active)* for 95 cents per minute; Extended Horoscope (900-370-2330) at $1.00 per minute; and School District data (900-GET-WISE), offering information on school districts on Long Island for $2.00 per minute.

Flat rate lines include Local Weather (900-443-5165) *(active)* for 75 cents per call; Gardening Tips with advice from columnist Margaret Roach (900-GROWING) for $1.00 per call; Best Bank Rates (900-GET-BEST) for $1.00 per call; Career Advice from columnist Patricia Kitchen (900-JOB-HELP) for $1.00 per call. *Feb. 1994*

Terrorism Protection Tips

"Guidelines For Protecting Yourself Against Terrorism" is a 900 line aimed at giving callers just that. For $2.49 per minute the line at 900-336-1248 offers basic tips that callers can use, such as, "Don't dismiss suspicious people."

IP Steve Gaul, who runs the line's parent company, SRV Services, with his wife and brother, said that the idea for the line, which began last Feb., seemed like a natural. The Gauls research material for the line with the FBI, the State Department and local police. *May 1994*

Top 14 Fax Info Calls List

The Detroit Free Press published a list of its Free Press Plus fax-on-demand All-Star team. These 14 documents are the most popular of those offered through the service.

Since its inception a year ago, readers' response to the service has been strong, said Rick Ratliff, associate director of Free Press Plus. "It seems to fill a gap for people in terms of information," he added. The service at 900-740-PLUS costs $2.50 per minute to order the document of a caller's choice.

The documents cover health, finance, law, taxes, government and real estate. Health concerns lead the list, with five health-related documents, four list approved mammography sites in different areas around Detroit, and one details breakthroughs in prostate cancer.

Other documents on the All-Star list include how to contact a member of Congress, IRS Form 942, instructions for Form 942, a checklist for preparing a home for sale, challenging an assessment increase, Social Security guidelines on household help, Mich. durable power of attorney form, Mich. living will form and Federal Debt/Deficit bibliography and chart. *May 1993*

Sleep Line

Callers to 1-900-73-SLEEP can hear recordings of relaxation and imagery techniques designed to help them fall asleep when insomnia strikes.

The $1.55 per minute line has an automatic disconnect feature after seven minutes, should the caller fall asleep before hanging up. The Sleepline recordings use relaxation and imagery techniques customized to offer relief from whatever may be keeping a caller awake, such as work-related stress or financial difficulties.
Sleep Media Network, Inc. sponsors the line. The service bureau is West Interactive Corp. *Sept. 1993*

UFO Info on 900

Just as a Menlo Park, California entrepreneur launched the now defunct UFO Line several years ago *(AN*, July 1990), UFO buff Glenn Segal of Mayfair, Pennsylvania began the UFO Information Line last December.

The $1.99-per-minute line at 900-884-UFOS offers callers a choice of two messages updated twice a month and averaging four minutes each. One message covers ongoing issues and past UFO encounters, and the other discusses latest sightings and subjects. Segal gathers information for the messages from publications on UFOs and declassified material from the government.

His marketing for the line includes ads in *Omni* magazine and the tabloid newspapers, *The National Enquirer* and *The Globe*.

AN cannot say what caused the cancellation of the earlier UFO line. In January of 1993 the line's IP complained to an *AN* editor that he experienced interference from the government who, he claimed wanted to hide information from the public. Segal compares the situation to an onion, saying, "You have to peel away the layers to find the truth of what the government is doing." *May 1994*

VETS System Still Going Strong

A 900 application with longevity is the VETS system run in partnership with the American Legion, Veterans of Foreign Wars (VFW) and GeoTel Corp. The system, begun in 1992 *(AN*, June '92), utilizes two separate 900 numbers for the American Legion and VFW, both aimed at providing veterans with dates and contact information for their military outfits' upcoming reunions.

Calls cost $1.95 per minute with the average call lasting two minutes. Live operators utilize a database of approximately 12,000 military reunions. Dick Ward, president of GeoTel Corp., said that the system logs 100,000 minutes per month, and the trend continues upward.

The VETS systems is marketed through listings in *American Legion* magazine, which has a circulation of three million, and *VFW*, which has a circulation of two million. Ward noted that the lines have peak periods for calls after each magazine comes out. *May 1994*

Yes & No About 900

How nice to be *The New York Times* and have the ability to advertise your own 900 number and refuse anyone else. On a recent Sunday, nestled in the much-read Broadway show listings of the Arts and Leisure section, was *The Time's* 900 number for article reprints.

At one article per call and $3.95 per article, the subjects of the six advertised reprints were healthy living & preventive care. Call 800-551-0159 for a faxed list of available articles on a variety of subjects. *Aug. 1993*

Phone Gambling

Phone Club International owns the rights to software designed for wagering via an 800 number. That number connects to an establishment in the Dominican Republic, where all wagers and money transactions occur electronically. The caller determines the amount of the wager ad the sporting event on which to gamble. The computer places the bet once the customer punches in a confirmation number.

Phone Club accounts reflect wins or losses. Members can receive money won by wire transfer or clearinghouse account.

Calling to place a bet is still illegal in the U.S., but according to FCC rules, calling a foreign country is not. Lawyers are still working out some details, but a spokesperson said the Phone Club will be up and running. A few other companies, like Sports Book International out of Antigua and Ladbrokes in the UK, operate similar operations. *October 1994*

TABS Offers 900
BBS Subscription Use

In its successful year-old service, TABS (Telephone Access Billing System), Palm Bay, Florida-based True Media Inc. offers the systems operators (or syops) for computer bulletin boards systems (BBS), 900 billing for their BBS subscribers.

When a BBS uses TABS, it gives out the two TABS 900 numbers to potential subscribers. TABS uses 900-622-8227 for BBS subscriptions for $10 worth of usage time, and 900-622-5225 for $25 subscriptions. During an approximately one-minute call, callers receive a subscription seven-digit ID number to use when dialing into the BBS. The $10 or $25 subscription charge shows up on their phone bills.

According to Rob Cartaino, True Media's director of marketing, 2300 BBS across the country currently use TABS. He noted that BBS are often limited in their billing capabilities because either they don't have the merchant accounts for credit cards or potential subscribers don't wish to give out their credit card numbers. "900 is a good option to get people to subscribe who normally wouldn't," Cartaino said.

True Media uses print media advertising in industry publications, such as *SYSOP News* and *CyberWorld Report* and *Board Watch Magazine* to market this billing system. Cartaino estimated that better than 50 percent of TABS users heard about the system through word-of-mouth from other sysops.

TABS users receive payment for each of their subscription I.D. numbers retrieved by callers to the 900 numbers. True Media deducts 25 percent for $10 subscriptions and 20 percent for $25 charges. Software and start-up is free to sysops. *Oct. 1994*

Appendix A
Magazines & Newsletters

Audiotex News. Published monthly by Audiotex News, Inc, 2362 Hampstead Turnpike, Second Floor, East Meadow, NY 11554
Telephone: 516-735-3398
Annual subscription: $249.00

This is the leading newsletter for the audiotex industry, published by Carol Morse Ginsburg, in a format which allows it to respond quickly to fast-breaking information and news about the audiotex industry. It does not accept advertising, and its purpose is "to give the audiotex industry access to the information it needs quickly, accurately, efficiently and to select and generate that information free from the influence of advertising." This newsletter is one of the few publications devoted exclusively to audiotex and pay-per-call, and is essential for the serious IP in keeping up-to-date about the industry.

InfoText. Published quarterly by Advanstar Communications, P.O. Box 6016, Duluth, MN 55806
Telephone: 800-346-0085 x-477
Annual subscription: $64.00

This magazine has gone through some changes over the years, starting as a monthly, then getting incorporated into *Voice Processing* magazine (which has itself since changed names to *Enterprise Communications*, see below), and is now published quarterly. This magazine is recognized as one of the major trade publications for the pay-per-call and interactive telephone industries.

InfoText contains current topical information relating to all facets of the audiotex industry, including new applications, legal up-dates, marketing information, and many newsworthy articles. This remains one of the best sources for information in the industry.

Enterprise Communications (formerly *Voice Processing Magazine*). Published monthly by Advanstar Communications, P.O. Box 6016, Duluth, MN 55806
Telephone: 800-346-0085 x-477
Annual subscription: $39.00

This magazine recently changed its title and expanded its scope to encompass unified messaging and information technologies. It continues to cover voice processing and information products, services, applications and technologies as well as voice mail, E-mail and interactive technologies. Its editorial mission is "To allow readers and vendors to exchange solutions to the challenges of integrating messaging and information exchange." This is the only trade magazine that is exclusively dedicated to unified messaging and the integration of voice, data and video information technologies.

Telemedia News and Views. Published monthly by OPUS Research, INC., 345 Chenery St., San Francisco, CA 94131
Telephone: 800-428-OPUS
Annual Subscription: $325

This newsletter was launched in May 1993 as the successor to *Audiotex Now* and *800/900 Review,* both of which were published by Strategic Telemedia, which has apparently decided to withdraw from the newsletter business. Nonetheless, OPUS Research is closely allied with Strategic Telemedia, and this newsletter is a vehicle for publishing some of Strategic Telemedia's research results. This newsletter is a good source for in-depth analysis and behind-the-scenes coverage of telemedia markets and trends, including new developments in the evolution of the information superhighway. An important source of timely information and intelligence for the serious players in this industry.

900 NewsReport. Published monthly by Moore Telecommunications, 6046 Cornerstone Ct., West, #126, San Diego, CA 92121
Telephone: 619-587-8126
Subscription: $35.00

Published by Toni Moore, a respected industry consultant and author of *Dialing For Dollars,* this newsletter offers helpful nuts-and-bolts advice aimed at the start-up IP, at an affordable price.

Teleconnect Magazine. Published monthly by Telecom Library, Inc., 12 West 21st St., New York, NY 10010
Telephone: 215-355-2886 (subscriptions)
          800-LIBRARY (publications & catalog)
Subscription: $15.00

This is one of the major trade magazines covering the overall telecommunications industry. According to its tag line, it is "the independent guide to choosing, using and installing telecommunications equipment and services." This magazine will keep you up-to-date on all facets of the telecom industry, of which audiotex is only a small part.

The Telecom Library also publishes *Call Center* magazine and several telecommunications books that may be of interest to you. Call or write for its catalog.

World Telemedia & Voice International. Both published by Triton Telecom Publishing, Ltd., 41-47 Kings Terrace, London NW1 0JR, United Kingdom
Telephone: 011 44 71 911 6002

These magazines are the international versions of *Infotext* and *Enterprise Communications,* respectively. Pay-per-call has already gone international, and these resources will help you keep on top of what's going on globally, primarily in Europe and Asia.

Phone+. Published monthly by Taurus Publishing, Inc., 4141 North Scottsdale Rd., Suite 316, Scottsdale AZ 85251
Telephone: 602-990-1101
Annual subscription: $50.00

This magazine is tag lined, "The Monthly Journal for thr Public Communications Industry," and is targeted to equipment and service providers in this industry, with an emphasis on long distance services and issues.

InformationWeek. Published weekly by CMP Publications, Inc., P.O. Box 1093, Skokie, IL 60076-8093
Telephone: 800-292-3642 ext. 40
Annual subscription: $63.95
This weekly magazine is written for information and technology managers and businesses -- or anyone who needs to stay up-to-date on what is going on on the information superhighway. Publisher CMP also publishes *CommunicationsWeek*, a sister magazine with more emphasis on equipment, technology and networks.

Interactive Age. Published biweekly by CMP Publications, Inc., (see above)
Annual subscription: $79.00
This biweekly newspaper covers the entire interactive industry, including the telecommunications, computer, entertainment/media and information industries, and how they are converging into a new interactive industry. Timely news and information for those who need to stay on top of developments in this fast-changing environment.

Telephony. Published weekly by Intertec Publishing, P.O. Box 12901, Overland Park, KS 66282-2901
Telephone: 800-441-0294
Annual Subscription: $45.00
This weekly magazine serves the public telephone network market, and its readers are primarily telecom professionals at the various local, regional and national telephone companies. Although the subject matter is targeted to these people, this is a good magazine for keeping current on what is happening with the carriers. *Telephony* lists *CommunicationsWeek* (see above) as one of its main competitors.

Appendix B
Trade Shows & Seminars

900 Business Seminars
Conducted by Carol Morse Ginsburg, editor & publisher of *Audiotex News*, 2362 Hempstead Tpke., 2nd Floor, East Meadow, NY 11554
Telephone: 516-735-3398
   These seminars are ideal for the start-up IP who wants a first-hand education by an experienced industry veteran. Seminars are offered regularly throughout the year in Hempstead, NY and Washington, DC. Call or write for the current schedule. If you cannot find the time to attend a seminar, cassette audio tapes of the full seminar are available for $49.95.

Audiotext Forum
R.j. Gordon & Company, Inc.
9200 Sunset Blvd., Suite 515, Los Angeles, CA 90069
Telephone: 310-278-8080   Fax: 310-274-8686
   With the demise of *InfoText's* annual conference in Las Vegas dedicated to the audiotext/pay-per-call industry, R.j. Gordon & Company has filled the gap with an event that specifically targets the information providers in this industry. Although the adult/entertainment side of the industry is heavily represented, this conference will be quite helpful for IPs offering any kind of information services. Like its *InfoText* predecessor, this event is held annually in January in Las Vegas. This is now the only trade show devoted exclusively to the audiotext and pay-per-call industries.

UMiiX (Unified Messaging & Interactive Information Exchange)
Business Communications Group, Advanstar Expositions,
201 E. Sandpointe Ave., Suite 600, Santa Ana, CA 92707
Telephone: 714-513-8645

This trade show is an important event for the voice processing/audiotext industry, replacing the *VOICE Exposition & Conference* and the *InfoText* sponsored exposition that was held each January in Las Vegas a few years ago. This trade show is a "forum in which users and suppliers of various means of messaging and information exchange can come together to discuss the issues that will affect enterprise-wide communications." It has a broader focus than the former *InfoText* and *Voice* shows, encompassing voice processing & automation, computer-telephone integration, and interactive information technologies, as well as audiotex and voice information applications.

According to its news release, "As companies continue to search for ways to integrate computer and telephone systems, UMiiX will focus on the convergence of voice, fax, e-mail, wireless and LAN-based communications, pulling all of these areas together into one understandable, solutions-oriented package."

If you can't afford the time or money to attend a trade show, purchasing the cassette tapes (seminar recordings) of the latest one is a good substitute. For more information call Advanstar Communications at 800-598-6008 or 216-243-8100.

Voice Asia/Europe/Mexico.
Advanstar Expositions, a Division of Advanstar Communications, Inc., P.O. Box 42382, Houston, TX 77242
Telephone: 713-974-6637   Fax: 713-974-6272

Advanstar Communications has gone international, with conferences held in London, Mexico and Hong Kong. These shows are focused on the trends, issues and opportunities for voice processing and audiotext manufacturers, service providers, and information providers in emerging markets.

World Telemedia Asia & Voice International Asia. Triton Telecom Publishing, Ltd., 41-47 Kings Terrace, London NW1 0JR, United Kingdom
Telephone: 011 +44 71 911 6002 Fax: 011 +44 71 911 6020

The pacific rim countries hold half the world's population, in economies that are growing fast. This Hong Kong exposition is for IPs who want to investigate Asia as a viable market. This is a two-track event, with World Telemedia covering audiotex and information services while Voice International covers the equipment side of the business.

Appendix C
Books & Directories

900 KNOW-HOW:
How to Succeed With Your Own 900 Number Business
By Robert Mastin
published by Aegis Publishing Group (1994)
796 Aquidneck Ave., Newport, RI 02842
Telephone: 401-849-4200   800-828-6961 Fax:401-849-4231
Price: $19.95

Now into its second edition, this information-packed 336 page book is widely recognized as the bible of the industry. It is written specifically for individuals or businesses that are planning to launch a 900 number information service. This is a nuts-and-bolts start-up guide with straightforward advice on what it will take to launch a successful program. The resource section consisting of 11 appendices (numbering more than 100 pages) is alone worth the cover price. This book is available directly from the publisher, or it can be found in most bookstores.

Marketing Your 900 Number: A User-Friendly Guide
By the editors of *Audiotex News*
published by *Audiotex News* (1994)
2362 Hempstead Tpke., 2nd Floor, East Meadow, NY 11554
Telephone: 516-735-3398   Fax: 516-735-3682
Price: $39.95

This is a one-of-a-kind reference guide packed with specific know-how and advice for marketing a 900 number. It covers all marketing options from getting free publicity with press releases to writing powerful ad copy to using infomercials. Also covers direct mail, media partners, TV advertising, radio and print advertising. This comprehensive guide covers the most critically important element of any successful 900 program: How to get your 900 number in front of the public and make those phones ring. Updated annually.

Opportunity is Calling:
How to Start Your Own Successful 900 Number
By Bob Bentz
Published by ATS Publishing (1993)
996 Old Eagle School Rd., Suite 1105, Wayne, PA 19087
Telephone: 610-688-6000
Price: $29.95

Written by Bob Bentz, the director of marketing at Advanced Telecom Services, one of the leading 900 service bureaus, this book is quite valuable for serious IPs in this business. In helping establish some 3,000 pay-per-call programs, Bob has probably seen every conceivable 900 application or idea, from the totally unworkable pie-in-the-sky scheme to the highly imaginative and well-conceived success story.

The Audiotex News Directory
By *Audiotext News* (see above)
Telephone: 516-735-3398
Price: $49.95

This is the only up-to-date comprehensive reference guide for the entire audiotex industry, including pay-per-call, automated voice response, computer telephony and voice information services. With more than 200 pages, if there's a telephone involved, this guide gives you the profiles and the detailed listings of the companies you're looking for. Features service bureaus, regulatory agencies, telephone companies, advertising agencies, hardware & software systems & components, publications, industry events and every other source you need. Published by *Audiotex News*, the leading newsletter for the pay-per-call industry, this directory must be on every player's desk.

Print Media Placement
By *Audiotex News* (see above)
Price: $49.95

Arranged alphabetically by state, this resource contains newspaper advertising rates, circulation, addresses, telephone numbers and restrictions (if any) for accepting 900 number advertising. Featured are more than 200 newspapers in 49 states, including dailies, weeklies and nationals. Updated annually.

The Power of 900
By Rick Parkhill
published by Advanstar Communications (1991)
7500 Old Oak Blvd., Cleveland, OH 44130
Telephone: 800-598-6008
Price: $45.00

This is the original book about the 900 industry. Great section on the history and evolution of the pay-per-call business. Some very useful directories and lists, including a directory of caller-paid services. The original edition was published in 1991 and has become somewhat dated, but a new edition may be forthcoming in the future.

Operating a "900" Number For Profit –
Entrepreneur Business Guide No. 1359
Published by Entrepreneur Group (1991)
2392 Morse Avenue, Irvine, CA 92714
Telephone: 714-261-2325
Price: $69.50

Published by the same company that publishes *Entrepreneur Magazine*, this is one of the latest in a long series of helpful business guides designed for the start-up entrepreneur. This comprehensive 200 page guide describes not only the specifics of starting a 900 business, but also the numerous considerations common to launching any new business venture. This guide will be particularly helpful for anyone who has never started a new business.

The Voice Response Reference Manual & Buyer's Guide
By Marc Robins
Published by Robins Press
2675 Henry Hudson Pkwy., West, Suite 6J, Riverdale, NY 10463
Price: $85.00
Telephone: 800-238-7130

This reference book is a complete resource for interactive voice technology, vendors and systems. It provides up-to-date information necessary for purchasing or building a voice response system. Included are comprehensive equipment vendor profiles and surveys on over 50 vendors and 60 systems. Hardware/software specifications, configurations & pricing, host computer interfaces, user management features, and selection/implementation advice are all included. Updated regularly.

DIALING FOR DOLLARS: A Guide to the 900 Business
By Antoinette Moore
Published by Moore Telecommunications Consultants (1993)
6046 Cornerstone Court West, Suite 126, San Diego, CA 92121
Telephone: 619-587-8126
Price: $29.95

Written by telecommunications consultant Toni Moore, this is a good introductory book about the 900/976 pay-per-call business, offering immediately useful nuts-and-bolts information for the start-up information provider. The step-by-step Activation Guide is both unique and helpful, and this book reflects the author's considerable experience in this business.

Newton's Telecom Dictionary
By Harry Newton
published by the Telecom Library, Inc.
12 West 21st St., New York, NY 10010
Telephone: 800-LIBRARY
Price: $24.95

This massive 1,120 page volume was written by Harry Newton, the publisher of *Teleconnect, Call Center* and *Imaging* magazines, in an easy-to-read non-technical style. This is an everyday working

---

dictionary for anyone involved in telecommunications. The user-friendly prose reads more like a good tutorial than a technical dictionary, and you'll never be confused again with the arcane language of the telecommunications industry.

The 900# Directory
Published by Pay Per Call Ventures (1993)
209 A Street, N.E., Washington, DC 20002
Telephone: 202-547-6595
Price: $129.00

This directory is a complete listing of over 700 newspapers that will accept 900 advertising, saving the advertiser considerable time in tracking down the right media for placing advertising.

The 900 Guide
By Madeline Bodin
published by the Telecom Library, Inc. (1993) (see above).
Price: $11.95

This helpful 96 page booklet offers sound advice from several experts in the 900 industry; including Brad Magill, Direct Response Broadcasting Network, Philadelphia, PA; Keith Dawson, associate editor of *Call Center* magazine; Deborah Vohasek, a voice response service bureau marketing expert; Gary Maier, president of Dianatel, a maker of PC boards; and Madeline Bodin, editor of *Call Center* magazine, to name only a few.

Telecom Made Easy:
Money-Saving, Profit-Building Solutions for Home Businesses, Telecommuters and Small Organizations
By June Langhoff
Published by Aegis Publishing Group (1995)
796 Aquidneck Ave., Newport, RI 02842
Telephone: 401-849-4200 800-828-6961 Fax: 401-849-4231
Price: $19.95

This is an easy-to-understand guide to getting the most out of telephone products and services. It is specifically written for small businesses, offices and organizations with fewer than 5 phone lines, who don't have telecom managers or the resources that big organizations have, but who still want to sound like them. Covers all the latest telecom products and services that are available, in simple, non-technical language, and how to put them together to best serve your specific needs. Available directly from the publisher or in most bookstores.

The McGraw-Hill TELECOMMUNICATIONS FACTBOOK
By Joseph A. Pecar, Roger J. O'Conner and David A. Garbin
Published by McGraw-Hill (1993)
Monterey Ave., Blue Ridge Summit, PA 17294
Telephone: 800-262-4729
Price: $29.95

This is a good guide for learning just about everything you will ever need to know about the overall telecommunications industry: Definitions, terminology, networks, LEC/IXC operations, analog vs. digital signals, transmission systems, circuit switching systems, premises distribution systems, and much more.

Service Access Codes 800/900 NXX Assignments (800/900 List)
Published quarterly by Bell Communication Research (Bellcore), and available by calling its Document Hotline:
Telephone: 800-521-2673
Price: $40.00 (subject to change)

This is a compilation of all NXXs assigned within the Service Access Codes (SACs) 800 and 900 by the North American Numbering Plan Administrator. For each of the NXXs assigned, the name of the company to which it is assigned and the telephone number of a contact in that company are given. NXX availability changes regularly, and this publication will give you the latest assignments.

Infopreneurs: Turning Data Into Dollars
By H. Skip Weitzen
published by John Wiley & Sons (1991)
605 Third Avenue, New York, NY 10158-0012
Price: $17.95

This book shows you how to make money selling information. Covers how to consolidate and communicate information; generate new information products and services; use volatile information profitably; price your computer services; and the use of the telephone and credit cards for instant payment.

PC-Based Voice Processing
By Bob Edgar
published by The Telecom Library, Inc. (1994)
12 West 21st St., New York, NY 10010
Telephone: 800-LIBRARY
Price: $34.95

This is the first book written exclusively about voice processing, aimed at the developer of a PC-based voice processing system. For programmers and value added resellers (VARs) in this growing industry.

Appendix D
Trade Associations

You will likely find numerous associations that can be of help to you in planning and researching your particular pay-per-call information service. Those associations that deal directly with the pay-per-call audiotext industry are listed here. There are many other associations that can also be of help, depending upon your circumstances, such as the Direct Marketing Assn. (DMA), the Promotion Marketing Assn. of America (PMAA), the U.S. Telephone Assn. (USTA) or the Telecommunications Industry Assn. (TIA). For a complete listing of associations consult the *Encyclopedia of Associations* by Gale Research, available in the reference section of larger libraries.

Interactive Services Association (ISA)
8403 Colesville Rd., Suite 865
Silver Spring, MD 20910
Telephone: 301-495-4955
Membership dues: Starting at $450
    The former National Association for Interactive Services (NAIS), which served the pay-per-call and audiotext industries for several years, was merged into the ISA in May 1994. The ISA is now the major trade association representing the interests of the evolving telecommunications-based personal interactive services industry.
    The ISA advocates the business and public policy interests of a wide variety of companies, including computer, consumer electronics,

publishing, telephone, software and other multimedia companies. Members include cable television companies, long distance carriers, RBOCs, audiotex companies, information providers, service bureaus, local exchange carriers, equipment manufacturers and a variety of others.

The ISA offers business development opportunities to its members through networking and cross-industry education. A sponsor of various conferences and member publications, the ISA provides a carefully screened stream of updated information to its members about the rapidly evolving interactive media industry.

Other important functions of the ISA are to advocate industry perspectives, to develop and enforce national industry standards through self-regulation, and to serve as a clearinghouse for industry information and education. Of particular value are the ISA's efforts in promoting a positive image for the industry and in lobbying with regulatory and legislative authorities with regard to proposed regulations and laws concerning the industry.

TeleServices Industry Association (TIA)
777 Alexander Rd., Suite 204
Princeton, NJ 08504
Telephone: 609-243-0066
Annual dues: Starting at $500

This association was established in January 1994 to serve the international audiotext industry. Membership is targeted to information providers in this industry, and the association's goal is to promote the welfare of its members by addressing their common goals and issues both locally and internationally.

The key roles of the TIA are to provide a forum to increase the dialog between the telephone companies, carriers, PTTs (foreign Post Telephone & Telegraph organizations), regulatory bodies and the rest of the audiotext industry and to promote a better understanding of the contributions made by all the participants. TIA also strives to develop and monitor a set of guidelines and procedures with the specific intent of comprehensively addressing the business policies from the perspectives of information providers, exchange carriers, and the consumers.

Information Industry Association (IIA)
555 New Jersey Avenue, N.W., Suite 800
Washington, DC 20001
Telephone: 202-639-8262
Membership fee: Starting at $500

IIA's Voice Information Services Division brings together equipment vendors, service bureaus, telecommunications companies, information providers, marketing organizations, consultants, and other businesses involved in the voice information field. Membership benefits are targeted to the larger players in the industry, and membership would be inappropriate for the start-up IP.

An established IP, on the other hand, should seriously consider joining this organization. IIA has been particularly active in pay-per-call consumer education, and has established Standards of Practice for voice information services.

Appendix E
Service Bureaus

This list of service bureaus was compiled in the fall of 1994, and was accurate at that time. Nonetheless, some of these companies may no longer be in business, or they may no longer serve 900 number information services.

The authors do not endorse any particular company in this list, and the fact that any given company is included in the list does not imply that this company is recommended in any way. You should follow prudent business practices in checking out any company you plan to do business with, including asking for and speaking with references.

See Appendix A, Magazines & Newsletters, for publications that offer service bureau listings that are kept more up-to-date than a book such as this can be. The authors regret any omissions or inappropriate inclusions, and invite service bureaus to contact us to make any necessary corrections or changes to this list for future editions of this book. Please call 401-849-4200 or write:

Aegis Publishing Group
796 Aquidneck Avenue
Newport, RI 02842-7202

This listing is divided geographically by state to facilitate locating a service bureau nearby. Nevertheless, geographic proximity is not critically important in most cases, and the capabilities of the bureau in serving your specific requirements is most important.

Alabama

Info Touch
Tim Brown, Vice President, Sales
3000 Zelda Rd. Suite F
Montgomery, AL 36106
205-244-9868

Information Management
Consultants (IMC)
Tim Michael, President
208 Adams St.
Mobile, AL 36603
205-434-6409 / 800-627-4IMC

Arizona

National Tel-Tec
Stewart Mazure, President
P.O. Box 4457
Scottsdale, AZ 85261
602-274-6444

Touch Tone Interactive, Inc.
Andrew Wise, President
Suite 6
637 North 3rd Ave.
Phoenix, AZ 85003
602-254-8250

California

Accelerated Voice
Jerry Nardini, President
25 Stillman St., Suite 200
San Francisco, CA 94107
415-543-2773

Alert Communications Co.
David Kissel, Sales Manager
5515 York Blvd.
Los Angeles, CA 90042
213-254-7171

Almarc
Ronald A. Resnick, President
8921 DeSoto, Suite 200
Conoga Park, CA 91306
818-773-2080

American International
Communications
Paul Keever
Suite 110
5595 East 7th St.
Long Beach, CA 90804
310-433-8818

Aspen Communications, Inc.
Barry Nadell, President
Suite 300
18757 Burbank Blvd.
Tarzana, CA 91356
818-774-2444

Automated Call Processing
Marcy McCann, Manager,
Business Development
244 Jackson St., Suite 200
San Francisco, CA 94111
415-989-2200

Bellatrix International
David Kahn, President
4055 Wilshire Blvd. Suite 415
Los Angeles, CA 90010
213-736-5600

Creative Call Management
Bob Kushner, President
316 W. 2nd St., Suite 1110
Los Angeles, CA 90012
213-687-0990

The Creative Services Group
Andy Batkin, President
2200 Pacific Coast Hwy., #103
Hermosa Beach, CA 90254
310-798-0433

DataDial International
Gordon Clements
650 Kenwyn Rd.
Oakland, CA 94610
510-601-0101

Dialtronix
Greg Samson, Sales Rep.
Suite 1100
4225 Executive Square
La Jolla, CA 92037
800-510-5500

E-Fax Communications, Inc.
(FAX Service Only)
William Perell, VP, Marketing
1611 Telegraph Ave., Suite 555
Oakland, CA 94612
510-836-6000

Elex Telemedia, Inc.
Larry Kerp
Suite 805
5777 West Century Blvd.
Los Angeles, CA 90045
310-568-4000

Gigaphone, Inc.
Nancy R. Conger, President
1525 Aviation Blvd., Suite A188
Redondo Beach, CA 90278
310-374-4313

Integrated Data Concepts
Warren Jason, President
P.O. Box 93428
Los Angeles, CA 90093
213-469-3380
800-367-4432

Interactive Strategies, Inc.
J. Edward Hastings, Ex. V.P.
31194 La Baya Dr., Suite 100
Westlake Village, CA 91362
818-879-9992

Intermedia Resources
Gene Chamson, President
6114 LaSalle Avenue, #230
Oakland, CA 94611
510-339-1792

Intertel Systems
Jeff Allen
P.O. Box 4384
Berkeley, CA 94704
510-649-0404

Kris Kupczyk & Assoc., Inc.
Kris Kupczyk
P.O. Box 8159
Calabasas, CA 91372
818-999-0644

Liaison 900
16821 San Fernando Mission
Blvd., Suite 402
Granada Hills, CA 91344
818-407-2727

Lorsch Creative Network, Inc.
Robert Lorsch, President
Suite 390
2934 1/2 Beverly Glen Circle
Los Angeles, CA 90077
310-476-6788

MCE TeleCommunications
Michaelle Ashlock
17911 Sky Park Circle, Suite D
Irvine, CA 92714
714-476-8007

Network Telephone
Services, Inc.
Gary Passon, President
6233 Variel Ave.
Woodland Hills, CA 91367
818-992-4300

New Media
Telecommunications, Inc.
Thomas Doolin, Sales Manager
Suite 1500
4225 Executive Square
La Jolla, CA 92037
619-558-3333

PhoneLink Group
12540 Beatrice St.
Los Angeles, CA 90066
310-301-4170

Resort Services Co.
Ann Pachter
Suite C
13912 Ponderosa St.
Santa Ana, CA 92705
714-848-0017

SimTel Communications
Jim Simpson, President
31220 LaBaya Dr., Suite 254
Westlake Village, CA 91362
818-706-1921

Speech Solutions
Tim Marentic, CEO
Suite 301
139 Townsend St.
San Francisco, CA 94107
415-243-8300

Strauss Communications, Inc.
Lance Strauss, President
P.O. Box 223542
Carmel, CA 93922
408-625-0700

Tel-Ad
Michael Newton
7760 E. Doheny Court
Anaheim Hills, CA 92808
714-281-1206

Telemedia Network
Steve Fecske
Suite 100
7551 Sunset Blvd.
Hollywood, CA 90046
213-845-1212

The Telephone Connection
Marc O'Krent, President
2554 Lincoln Blvd., Suite 137
Marina del Rey, CA 90291
310-827-8787

Teleserve, Inc.
Sales Department
47000 Warm Springs Blvd.,
#460
Fremont, CA 94539
408-727-7764

TeleTel, Inc.
Horace Zhang
100 Wilshire Blvd., Suite 420
Santa Monica, CA 90401
310-458-6333

Unlimited Telesources
Jacquline Shaw
Suite 3
1134 23rd St.
Santa Monica, CA 90403
310-458-1506

US Teleconnect
Dana Dunn
Suite B
432 Bonito Ave.
Long Beach, CA 90802
310-436-1326

WICE Telecommunications
R. Varasteh
17911 Sky Park Circle
Irvine, CA 92714
714-476-8007

Colorado

Advanced Communications Svcs
Jerry Krenning
155 N. College Ave., #222
Ft. Collins, CO 80524
303-482-5542

Cook Communications Corp.
Jim Moreland, Director of Sales
Genesee Ctr. 1 , Suite 120
602 Park Point Dr.
Denver, CO 80401
800-FON-CALL/303-526-7400

IdealDial
Michael Couglin, Vice President
910 15th St., Suite 900
Denver, CO 80202
800-582-3425

Interactive Information Systems
Paul Kulas, Vice President
910 15th St., Suite 751
Denver, CO 80202
303-595-0888

Tela, Inc.
Ted M. Larson, President
910 15th St., Suite 1068
Denver, CO 80202
303-893-5150

Connecticut

Facsimile Marketing, Inc.
Jeremy Grayzel, President
3 Landmark Square, Suite 403
Stamford, CT 06901
203-323-4400

VIP Communications
Ken Hovland, Jr., President
P.O. Box 180

130 Old Town Rd.
Vernon, CT 06066
203-875-3366

Delaware

American TelNet, Inc.
William Rivell, VP, Sales
1701 Augustine Cut-Off, Suite 40
Wilmington, DE 19803
302-651-9400

Mega900 Communications
Jeff Platt
901 West 24th St.
Wilmington, DE 19802
302-654-0829

District
of Columbia

Telecompute Corp.
Warren Miller, President
1275 K St., NW, Suite G-9
Washington, DC 20005
202-789-1111
800-872-8648

Florida

Alternative Communications
& Technologies, Inc.
Barbara Johnson
12807 W. Hillsborough Ave., # J
Tampa, FL 33635
813-854-1755

Audio Text Ventures, Inc.
Joni Lawler
200 West Forsythe St., Ste. 800
Jacksonville, FL 32202
904-346-1303

Audiotext Services, Inc.
Roderic van Beuzekom,
President
P.O. Box 2449
Orlando, FL 32802
407-426-8355

Avalon Communications
1007 North Federal Highway
Sales Department
Ft. Lauderdale, FL 33304
800-391-6555

Connections USA, Inc.
Yvette Momot
P.O. Box 030459
Ft. Lauderdale, FL 33303-04
305-525-4141

Florida Group Services
Lee Watson, Dir. of Marketing
Suite 310
2500 Maitland Center Pkwy.
Maitland, FL 32751
800-432-4347

HSN 800/900 Corporation
Bill Clark
P.O. Box 9090
Clearwater, FL 34618
813-572-8585

ICN Corporation
Eric Bachkoff, Sales Director
Suite 300
1801 South Federal Hwy.
Delray Beach, FL 33483
407-272-5667

Megatech Services
Jack Potenza
2308 23rd Way
West Palm Beach, FL 33407
407-689-7493

ONE 800/900, Inc.
Cynthia King, Sales Dept.
200 Laura St. 12th Floor
Jacksonville, FL 32202
904-355-9000

Phoneworks
Kathy Montgomery
146 Second St. N., Suite 201
St. Petersburg, FL 33701
813-823-7144

Saturn Communications
Mark Amarant
P.O. Box 848367
Hollywood, FL
305-438-7000

Georgia

The Intermedia Group
Steven West
Suite 500, Bl. 192
2980 Cobb Pkwy.
Atlanta, GA 30339
404-368-2838

Message Technologies, Inc.
Mark Abramson, President
2849 Paces Ferry Rd., #600
Atlanta, GA 30339
800-868-3684

Overlook Communications
International
Jan Hart
2839 Paces Ferry Rd., Suite 500
Atlanta, GA 30339
404-432-6800

Tamona Enterprises, Inc.
Pati Johnston
8215 Roswell Rd., Bldg. 900
Atlanta, GA 30350
800-356-4466/404-604-3500

Technology Solutions
International
Joe Rosenthal, President
1400 Lake Hern Dr. NE, Ste 170
Atlanta, GA 30319
404-843-5890

Illinois

Ameritech Audiotex Services, Inc.
Jeff Thadie, Dir., Sales & Mktg.
300 South Riverside Plaza
Chicago, IL 60606
312-906-3130

Document Retrieval Services
Ronald Duskey, President
644 W. Pratt Ave. North
Schaumburg, IL 60193
708-924-7464

DynamicFax, Inc.
James L. Hughes, VP, Sales
2470 Eastrock Dr.
Rockford, IL 61108
815-398-9009

Information Command, Inc.
Donald Young, President
444 North Wells
Chicago, IL 60610
312-245-1111/800-282-6541

LO/AD International
George Kois
1361 Champion Forest Ct.
Wheaton, IL 60187
708-668-8788

New Tech Telemedia
John Wellbourn, VP Sales
444 N. Wells St.
Chicago, IL 60610
312-245-1111

Northwest Nevada Telco
Michael Dawson, Dir. of Sales
1324 Evers Ave.
Westchester, IL 60154-3413
800-279-0909

Stargate Communications, Inc.
G. Gary Chaffin
628 Gannet Ln.
Bolingbrook, IL 60440
800-282-6541

Iowa

MCI Information Resources
Renee Monteric
500 2nd Ave., SE
Cedar Rapids, IA 52403
319-366-6600

Kansas

Brite Voice Systems, Inc.
Carolyn Russell, Director,
Marketing Communications
7309 E. 21st. St. North
Wichita, KS 67206
800-SEE-BRITE (sales)
316-652-6500

Info Access, Inc.
Terry Hughes, Sales Director
4550 West 109th St.
Overland Park, KS 66211
800-453-1453

Maryland

National Phone Link
Scott Kleinknecht, President
6900 Virginia Manor, Suite 110
Beltsville, MD 20705
301-419-0365

Order by Phone, Inc.
Michael Metzger, President
1 East Chase St., Ste. 200
Baltimore, MD 21202
410-837-0800

TeleService USA
Barry Cockley, Executive VP
19723 Leitersburg Pike
Hagerstown, MD 21742
301-797-2323

Telesonic
E. Escobar, Sales Representative
120 Admiral Cochrane Dr.
Annapolis Science Center
Annapolis, MD 21401
410-841-6920

Massachusetts

Facsimile Services, Inc.
Edward Olkkola, President
573 Washington St.
South Easton, MA 02375
508-230-2000

Inpho, Inc.
Steve Kropper, President
225 Fifth St.
Cambridge, MA 02142
617-868-7050

Instant Information
Brian Cavoli, Project Manager
5 Broad St.
Boston, MA 02109
617-523-7636

Mass Communication
Stephen R. Picardo, President
432 Columbia St., Suite B-9
Cambridge, MA 02141
617-577-7285

Pilgrim Telephone, Inc.
David Silver
Suite 104
719 Washington St.
Newtonville, MA 02160
617-577-7575

Tele-Publishing, Inc.
Paul Twitchell,
Director of Marketing
126 Brookline Ave.
Boston, MA 02215
617-536-2340
800-874-2340

Michigan

Amrigon
Michelle Gustavus,
Account Representative
2750 South Woodward
Bloomfield Hills, MI 48304
313-332-2300

Dobbs Enterprises, Ltd.
John C. Dobbs, President
855 Forest St.
Birmingham, MI 48009
313-540-2149

World Data Delivery Systems
Matt Kennedy, Telcom Director
20542 Harper Ave.
Harper Woods, MI 48225
800-554-9337

Minnesota

Micro Voice Applications, Inc.
Michael A. James
5775 Wayzata Blvd.
Minneapolis, MN 55416
800-553-0003

Talk, Inc.
Patrick Dolan, Vice President
3415 University Ave., SE
Minneapolis, MN 55414
612-642-4559
612-642-4558

Missouri

Available Communications, Inc.
Suite 106
11330 Olive Blvd.
St. Louis, MO 63141
314-995-9000

Nebraska

Call Interactive
Deanna L. DeSmet,
Director of Marketing
2301 North 117th Ave.
Omaha, NE 68164
800-428-2400

Prairie Systems
Christine Reilly
Suite 310
9300 Underwood Ave.
Omaha, NE 68114
402-398-4121

SITEL Corp.
Kevin Blair, VP, Advertising
5601 N. 103rd St.
Omaha, NE 68134
402-498-6876/402-498-6810
800-445-6600

WATS Marketing
Joseph Duryea
2121 North 117th Ave.
Omaha, NE 68164
800-351-1000

Wessan Interactive Network
Karen Westerfield, President
3033 North 93rd St.
Omaha, NE 68134
800-468-7800

West Interactive Corp.
David VanDerveer, Executive VP,
Sales & Marketing
9223 Bedford Ave.
Omaha, NE 68134
800-841-9000

Nevada

Arch Communications
Steve Carne
Suite1000
3700 S. Las Vegas Blvd.
Las Vegas, NV 89109
702-597-1829

Audio Communications, Inc.
John Wennstrom,
General Manager
Suite 24
3110 Polaris Ave
Las Vegas, NV 89102
800-527-5353

B.F.D. Productions
Bruce F. Dyer, President
1210 S. Martin Luther King
Las Vegas, NV 89102
702-387-3200
800-444-4BFD

Technology Support Corporation
Ben Greenspan, President
PO Box 6494
Incline Village, NV 89450
702-832-7358

Appendix E

New Hampshire

Tolles Communications
Irv Tolles, President
103 Bay St., P.O. Box 240
Manchester, NH 03105
800-747-1667

New Jersey

900 Plus Communications
Al Edward
25 Main St., Crt. Plaza N.
Hackensack, NJ 07601
221-489-8909

American Audio Service Bureau
Betty Fogal
2007 Hoyt Ave., 2nd Floor
Fort Lee, NJ 07024
201-585-8100

Coopers & Lybrand
Dan Ferreiza
1 Sylvan Way
Parsippany, NJ 07054
201-829-9520

Evergreen Systems, Inc.
Michael Osborn
PO Box 1400
Marlton, NJ 08053
609-985-3333

The Fax People, Inc.
Michael Rogers, President
34 Maple St., Suite 3
Summit, NJ 07901
800-FAX-NET1/908-277-2122

The Interactive Telephone Co.
Abby Knowlton, Sales Director
25 Main St., 2nd Floor
Hackensack, NJ 07601
201-342-1000

International Teleprograms
David Fritz
Suite 1225
24 Commerce St.
Newark, NJ 07102
201-624-6020

Most Telecom, Inc.
Ron Patetta, VP, Sales
18 Condit Rd.
Mountain Lakes, NJ 07046
210-335-2255

RENEW Interactive
Marketing Services, Inc.
Henry Wener, President
15 Monhegan St.
Clifton, NJ 07013-2009
201-614-0171

Sigma Communications
Services, Inc.
Michael Rogers, President
Suite 203
350 Springfield Ave.
Summit, NJ 07901
908-277-2122

Voice Communications, Inc.
Frank Giarratano
350 Main Rd.
Montville, NJ 07045
201-299-1200

New York

Audio-Voice, Inc.
Jo-Ann Sickinger
Suite 401
545 8th Ave.
New York, NY 10018
212-868-1121

Automated Fax
& Voice Solutions
Burt Anderson, President
90 Merrick Ave.
East Meadow, NY 11554
516-674-4600
516-794-1100

Eagle Communications, Inc.
James Park, Dir. of Marketing
127 East 59th St., 2nd Floor
New York, NY 10022
212-758-7744

Fax Information
Network of America, Inc.
Ralph Potente, President
20 Max Ave.
Hicksville, NY 11801
516-942-8000

Integrated Communications Ltd.
Jonathan Breger, Vice President
885 Third Ave., Suite 330
New York, NY 10022
212-230-2222

Music Access
Bar Bizick
90 Fifth Ave.
Brooklyn, NY 11217
718-398-2146

Peak Communications
David Peterson, President
41 East 60th St.
New York, NY 10022
212-980-1949

Phone Programs, Inc.
Carol Brennon
40 Elmont Rd.
Elmont, NY 11003
516-775-5410

TRX Corp.
Gary Glicker
160 East 56th St.
New York, NY 10022
212-644-0370

VOCALL Communications Corp.
Laura Pettinato
70 East 55th St., Heron Tower
New York, NY 10022
212-754-2525

Voice One
James Lawyer, President
P.O. Box 13233
Albany, NY 12212
518-434-6499

Weather Concepts, Inc.
Tore Jakobsen
1966 Route 52
Hopewell Junction, NY 12540
914-226-8200

Ohio

900 America, Ltd.
Larry D. Lomaz, CEO
1 Cascade Plaza, Suite 1940
Akron, OH 44308
216-379-9900

Digital Communications
R.J. Jones
P.O. Box 3212
Youngstown, OH 44513
800-769-3238

ITI Marketing Services
Kirk VonDerhaar
531 North Wayne Ave.
Cincinnati, OH 45215
513-563-8666/800-562-5000

Scherers Communications, Inc.
MaryKay Dawson
575 Scherers Ct.
Worthington, OH 43085
614-847-6161

Worthington Voice Services, Inc.
De Trinh
Suite E
740 East Lakeview Plaza
Worthington, OH 43085
614-431-9710

Oklahoma

900 Call Association
Thomas Hoshall, President
3608 NW 58th St.
Oklahoma City, OK 73112
405-947-5627

VoiceXpress, Inc.
Kevin Murray, Vice President
15 West 6th St., Suite 1310
Tulsa, OK 74119
918-583-8080

Oregon

Sonnet Communications
Diane Ulicsni
1075 NW Murray Rd., Ste. 262
Portland, OR 97229
503-233-3635

Pennsylvania

Accu-Weather, Inc.
Sheldon Levine, Director of Sales
619 W. College Ave.
State College, PA 16801
814-237-0309

Advanced Telecom Services, Inc.
Bob Bentz, Director of Marketing
996 Old Eagle School Rd.
Wayne, PA 19087
610-688-6000

Automated Voice
Production, Inc.
Gary Baron
295 Buck Rd., Suite 207
Holland, PA 18966
215-953-8568

General Fax, Inc.
Christopher Stephano, President
Suite A103
Front & Ford Sts.
Bridgeport, PA 19405
215-277-1722

Gralin Associates, Inc.
Account Representative,
3605A Old Easton Rd.
Doyleston, PA 18901
800-724-0992

Inter#Net, Inc.
Barry Krueger, Director of Sales
1001 East Entry Dr., Suite 110
Pittsburgh, PA 15216
412-571-3350

Sports Network
Ken Zajac
701 Masons Mill Business Park
Huntington Valley, PA 19006
215-947-2400

Voice FX Corp.
Chris Gongol, VP, Marketing
4th Floor
1100 E. Hector St.
Conshohocken, PA 19428
215-941-1000

South Carolina

Info-Tel
Chris Brunson
217 Lucas St., Suite E
Mt. Pleasant, SC 29464
800-388-3528

Texas

Action Fax International, Inc.
Jim Hawkins, VP Sales
4851 Keller Springs Rd.
Dallas, TX 75248
214-931-5800

Celebration Computer Systems
James F. Wiseheart, President
9207 Country Creek, Suite 140
Houston, TX 77036
713-625-4000

CommNet, Inc.
Dave Knapp, President
1206 Avenue R, Suite D
Lubbock, TX 79401
806-747-3025

Fax Access Xchange, Inc.
(FAX, Inc.)
Jerry Bachmann, VP, Sales
4851 Keller Springs Rd.
Dallas, TX 75248-5928
214-931-5800

Modern Moral Innovations (MMI)
Jay Dietz, President
4019 Adonis
Houston, TX 77373
713-350-4664

NeoData
Kevin McKinnon, Marketing Mgr.
100 Crescent Ct., Suite 650

Dallas, TX 75201
214-871-5588

SABRE Teleservice Resources
Donna Webb, Sales Manager
4201 Cambridge Rd.
Ft. Worth, TX 76155
800-352-2580

Voicetext Interactive
Eileen Williams, President
702 Colorado St., Suite 125
Austin, TX 78701
512-404-2300

WTS Bureau Systems, Inc.
David Matthews
2170 Lone Star Dr.
Dallas, TX 75212
214-920-1900

Utah

Co-op Communications
3227 N. Canyon Rd.
Provo, UT 84604
801-224-8366

Teleshare
Shane Heath, Dir. of Operations
Suite 103
227 North University Ave.
Provo, UT 84601
801-377-0600

Virginia

Phone Base Systems, Inc.
Phil Gross, Senior Vice President
8620 Westwood Center Dr.
Vienna, VA 22182
703-893-8600

Ultra Communications
Paul Dalton, President
4330 M Evergreen Lane
Annandale, VA 22003-3211
703-642-3100

Washington

Automotive Experts
Tom Barrett
P.O. Box 2994
Renton, WA 98056
206-271-7200

Bureau One, Inc.
Chris Schott, Director of New
Accounts
921 14th St.
Longview, WA 98632
206-636-2000

Pick-A-Winner, Inc.
John Jordan
10020 Main St., Ste. A4
Bellevue, WA 98004-6000
206-453-4800

Wisconsin

Internet, Inc.
Bill Luebke
741 N. Milwaukee St., #420
Milwaukee, WI 53202
414-274-3820

Canada

Corporate Telemarketing, Inc.
David Filwood, President
12 Water St., Suite 401
Vancouver, BC V6B 1A5
604-685-6144

Fax Canada
Alana Samuels, President
980 Alness St., Unit 13
North York, Ontario M3J 2S2
416-650-9314

Interactive Telephone Marketing
822 Richmond St. W., Suite 200
Toronto, Ontario M6J 1C9
416-363-5000

Messagebank of Calgary
Maury Wasserman, President
222 58th Ave. SW, #212
Calgary, Alberta T2H 2S3
403-531-4000

Phoneworks Canada
Dan Melymuk
181 Carlaw Ave., Suite 310
Toronto, Ontario M4M 2S1
416-778-7877

PRIMA Telmatic Inc.
14 Commerce Pl., Suite 510
Nuns' Island, Quebec H3E 1P5
514-768-7676

Voice Courrier
Gordon Black, Sales Manager
1300 Bay St., Suite 200
Toronto, Ontario M5R 3K8
416-921-0033

Interactive Entertainment Group
Darian Brooks, President
26 Parkwood Ave.
Toronto, Ontario M4V 2X1
800-567-8965

Telepath Telecommunications
Warren Eugene, President
26 Parkwood Ave.
Toronto, Ontario M4V 2X1
800-567-8965

Appendix F
Federal Laws & Regulations

Federal Communications Commission (FCC) Regulations

The FCC regulations that resulted from passage of the Telephone Disclosure and Dispute Resolution Act (TDDRA), the relevant federal law, are reproduced in full here. The following is a verbatim transcription from Title 47 of the Code of Federal Regulations, Part 64, Sections 64.1501 through 64.1515, which are the complete FCC regulations governing pay-per-call services:

Section 64.1501 Definitions

For the purposes of this subpart, the following definitions shall apply:
(a) Pay-per call service means any service
    (1) In which any person provides or purports to provide
        (A) Audio information or audio entertainment produced or packaged by such person;
        (B) Access to simultaneous voice conversation services; or
        (C) Any service, including the provision of a product, the charges for which are assessed on the basis of the completion of the call;
    (2) For which the caller pays a per-call or per-time-interval charge that is greater than, or in addition to, the charge for transmission of the call; and
    (3) Which is accessed through use of a 900 telephone number.
(b) Such term does not include directory services provided by a common carrier or its affiliate or by a local exchange carrier or its affiliate, or any service the charge for which is tariffed, or any service for which users are assessed charges only after

entering into a presubscription or comparable arrangement with the provider of such service.

(b)(1) Presubscription or comparable arrangement means a contractual agreement in which

(i) The service provider clearly and conspicuously discloses to the consumer all material terms and conditions associated with the use of the service, including the service provider's name and address, a business telephone number which the consumer may use to obtain additional information or to register a complaint, and the rates for the service;

(ii) The service provider agrees to notify the consumer of any future rate changes;

(iii) The consumer agrees to utilize the service on the terms and conditions disclosed by the service provider; and

(iv) The service provider requires the use of an identification number or other means to prevent unauthorized access to the service by nonsubscribers.

(2) Disclosure of a credit or charge card number, along with authorization to bill that number, made during the course of a call to an information service shall constitute a presubscription or comparable arrangement if the credit or charge card is subject to the dispute resolution procedures of the Truth in Lending Act and Fair Credit Billing Ace, as amended, 15 U.S.C. Section 1601 et seq. No other action taken by the consumer during the course of a call to an information service, for which charges are assessed, can be construed as creating a presubscription or comparable arrangement.

64.1502 Limitations on the Provision of Pay-Per-Call Services

Any common carrier assigning a telephone number to a provider of interstate pay-per-call service shall require, by contract or tariff, that such provider comply with the provisions of this subpart and of titles II and III of the Telephone Disclosure and Dispute Resolution Act (Pub. L. No. 102-556) (TDDRA) and the regulations prescribed by the Federal Trade Commission pursuant to those titles.

64.1503 Termination of Pay-Per-Call Programs.

Any common carrier assigning a telephone number to a provider of interstate pay-per-call service shall specify by contract or tariff that pay-per-call programs not in compliance with Section 64.1502 shall be terminated following written notice to the information provider. The information provider shall be afforded a period of no less than seven and no more than 14 days during which a program may be brought into compliance. Programs not in compliance at the expiration of such period shall be terminated immediately.

64.1504 Restrictions on the Use of 800 Numbers.

Common carriers shall prohibit, by tariff or contract, the use of any telephone number beginning with an 800 service access code, or any other telephone number advertised or widely understood to be toll free, in a manner that would result in

(a) The calling party or the subscriber to the originating line being assessed, by virtue of completing the call, a charge for the call;

(b) The calling party being connected to a pay-per-call service;

(c) The calling party being charged for information conveyed during the call unless the calling party has a presubscription or comparable arrangement; or

(d) The calling party being called back collect for the provision of audio or data information services, simultaneous voice conversation services, or products.

64.1505 Restrictions on Collect Telephone Calls.

(a) No common carrier shall provide interstate transmission or billing and collection services to an entity offering any service within the scope of 64.1501 (a)(1) that is billed to a subscriber on a collect basis at a per-call or per-time-interval charge that is greater than or in addition to, the charge for transmission of the call.

(b) No common carrier shall provide interstate transmission services for any collect information services billed to a subscriber at a tariffed rate unless the called party has taken affirmative action clearly indicating that it accepts the charges for the collect service.

64.1506 Number Designation.

Any interstate service described in 64.1501 (a)(1) - (2) shall be offered only through telephone numbers beginning with a 900 service access code.

64.1507 Prohibition on Disconnection or Interruption of Service for Failure to Remit Pay-Per-Call or Similar Service Charges.

No common carrier shall disconnect or interrupt in any manner, or order the disconnection or interruption of, a telephone subscriber's local exchange or long distance telephone service as a result of that subscriber's failure to pay

(a) Charges for interstate pay-per-call service,

(b) Charges for interstate information services provided pursuant to a presubscription or comparable arrangement or

(c) Charges, which have been disputed by the subscriber, for interstate tariffed collect information services.

64.1508 Blocking Access to 900 Service

(a) Local exchange carriers must offer to their subscribers, where technically feasible, an option to block access to services offered on the 900 service access code. Blocking is to be offered at no charge, on a one-time basis, to

(1) All telephone subscribers during the period from November 1, 1993 through December 31, 1993; and

(2) Any subscriber who subscribes to a new telephone number for a period of 60 days after the new number is effective.

(b) For blocking requests not within the one-time option or outside the time frames specified in paragraph (a) of this section, and for unblocking requests, local exchange carriers may charge a reasonable one-time fee. Requests by subscribers to remove 900 services blocking must be in writing.

(c) The terms and conditions under which subscribers may obtain 900 services blocking are to be included in tariffs filed with this Commission.

64.1509 Disclosure and Dissemination of Pay-Per-Call Information.

(a) Any common carrier assigning a telephone number to a provider of interstate pay-per-call services shall make readily available, at no charge, to Federal and State agencies and all other interested persons:

(1) A list of the telephone numbers for each of the pay-per-call services it carries;

(2) A short description of each such service;

(3) A statement of the total cost or the cost per minute and any other fees for each such service; and

(4) A statement of the pay-per-call service provider's name, business address, and business telephone number.

(b) Any common carrier assigning a telephone number to a provider of interstate pay-per-call services and offering billing and collection services to such provider shall

(1) Establish a local or toll-free telephone number to answer questions and provide information on subscribers' rights and obligations with regard to their use of pay-per-call services and to provide to callers the name and mailing address of any provider of pay-per-call services offered by that carrier; and

(2) Provide to all its telephone subscribers, either directly or through contract with any local exchange carrier providing billing and collection services to that carrier, a disclosure statement setting forth all rights and obligations of the subscriber and the carrier with respect to the use and payment of pay-per-call services. Such statement must include the prohibition against disconnection of basic communications services for failure to pay-per-call charges established by 64.1507, the right of a subscriber to obtain blocking in accordance with 64.1508, the right of a subscriber not to be billed for pay-per-call services not offered in compliance with federal laws and regulations established by 64.1510 (a) (iv), and the possibility that a subscriber's access to 900 services may be involuntarily blocked pursuant to 64.1512 for failure to pay legitimate pay-per-call charges. Disclosure statements must be forwarded to:

(i) All telephone subscribers no later 60 days after these regulations take effect;

(ii) All new telephone subscribers no later than 60 days after service is established;

(iii) All telephone subscribers requesting service at a new location no later than 60 days after service is established; and

(iv) Thereafter, to all subscribers at least once per calendar year, at intervals of not less than 6 months nor more than 18 months.

64.1510 Billing and Collection of Pay-Per-Call and Similar Service Charges.

(a) Any common carrier assigning a telephone number to a provider of interstate pay-per-call services and offering billing and collection services to such provider shall:

(1) Ensure that a subscriber is not billed for interstate pay-per-call services that such carrier knows or reasonably should know were provided in violation of the regulations set forth in this subpart of prescribed by the Federal Trade Commission pursuant to titles II or III of the TDDRA or any other federal law;

(2) In any billing to telephone subscribers that includes charges for any interstate pay-per-call service

(i) Include a statement indicating that:

(A) Such charges are for non-communications services;

(B) Neither local nor long distances services can be disconnected for non-payment although an information provider may employ private entities to seek to collect such charges;

(C) 900 number blocking is available upon request; and

(D) Access to pay-per-call services may be involuntarily blocked for failure to pay legitimate charges;

(ii) Display any charges for pay-per-call services in a part of the bill that is identified as not being related to local and long distance telephone charges;

(iii) Specify, for each pay-per-call charge made, the type of service, the amount of the charge, and the date, time and, for calls billed on a time-sensitive basis, the duration of the call; and

(iv) Identify the local or toll-free number established in accordance with 64.1509 (b)(1).

(b) Any common carrier offering billing and collection services to an entity providing interstate information services pursuant to a presubscription or comparable arrangement, or for interstate tariffed collect information services, shall, to the extent possible, display the billing information in the manner described in paragraphs (a) (2) (i) - (ii) of this section.

64.1511 Forgiveness of Charges and Refunds.

(a) Any carrier assigning a telephone number to a provider of interstate pay-per-call services or providing transmission for interstate tariffed collect information services or interstate information services offered under a presubscription or comparable arrangement, and providing billing and collection services for such services, shall establish procedures for the handling of subscriber complaints regarding charges for those services. A billing carrier is afforded discretion to set standards for determining when a subscriber's complaint warrants forgiveness, refund or credit of interstate pay-per-call or information services charges provided that such charges must be forgiven, refunded, or credited when a subscriber has complained about such charges and either this Commission, the Federal Trade Commission, or a court of competent jurisdiction has found or the carrier has determined, upon investigation, that the service has been offered in violation of federal law or the regulations that are either set forth in this subpart or prescribed by the Federal Trade Commission pursuant to titles II or III of the TDDRA. Carriers shall observe the record retention requirements set forth in 47 C.F.R. Section 42.6 except that relevant records shall be retained by carriers beyond the requirements of Part 42 of this chapter when a complaint is pending at the time the specified retention period expires.

(b) Any carrier assigning a telephone number to a provider of interstate pay-per-call services but not providing billing and collection services for such services, shall, by tariff or contract, require that the provider and/or its billing and collection agents have in place procedures whereby, upon complaint, pay-per-call charges may be forgiven, refunded, or credited, provided that such charges must be forgiven, refunded, or credited when a subscriber has complained about such charges and either this Commission, the Federal Trade Commission, or a court of competent jurisdiction has found or the carrier has determined, upon investigation, that the service has been offered in violation of federal law or the regulations that are either set forth in this subpart or prescribed by the Federal Trade Commission pursuant to titles II or III of the TDDRA.

64.1512 Involuntary Blocking of Pay-Per-Call Services

Nothing in this subpart shall preclude a common carrier or information provider from blocking or ordering the blocking of its interstate pay-per-call programs from numbers assigned to subscribers who have incurred, but not paid, legitimate pay-per-call charges, except that a subscriber who has filed a complaint regarding a particular pay-per-call program pursuant to procedures established by the Federal Trade Commission under title III of the TDDRA shall not be involuntarily blocked from access to that program while such a complaint is pending. This restriction is not intended to preclude involuntary blocking when a carrier or IP has decided in one instance to sustain charges against a subscriber but that subscriber files additional separate complaints.

64.1513 Verification of Charitable Status.

Any common carrier assigning a telephone number to a provider of interstate pay-per-call services that the carrier knows or reasonably should know is engaged in soliciting charitable contributions shall obtain verification that the entity or individual for whom contributions are solicited has been granted tax exempt status by the Internal Revenue Service.

64.1514 Generation of Signalling Tones.

No common carrier shall assign a telephone number for any pay-per-call service that employs broadcast advertising which generates the audible tones necessary to complete a call to a pay-per-call service.

64.1515 Recovery of Costs.

No common carrier shall recover its cost of complying with the provisions of this subpart from local or long distance ratepayers.

FCC Dial-a-Porn Rules

Because the FCC Dial-a-Porn Rules apply to the use of indecent communications over all telephone lines, not just 900 pay-per-call lines, these rules are distinct from the FCC 900 Rules that were just presented. The following is a verbatim transcription of the FCC Dial-a-Porn Rules:

64.201 Restrictions on Indecent Telephone Message Services

(a) It is a defense to prosecution for the provision of indecent communications under section 223(b)(2) of the Communications Act of 1934, as amended (the Act), 47 U.S.C. 223(B)(2), that the defendant has taken the action set forth in paragraph (a)(1) of this section and, in addition, has complied with the following: Taken one of the actions set forth in paragraphs (a)(2), (3), or (4) of this section to restrict access to prohibited communications to persons eighteen years of age or older, and has additionally complied with paragraph (a)(5) of this section, where applicable:

(1) Has notified the common carrier identified in section 223(c)(1) of the Act, in writing, that he or she is providing the kind of service described in section 223(b)(2) of the Act.

(2) Requires payment by credit card before transmission of the message; or

(3) Requires an authorized access or identification code before transmission of the message, and where the defendant has:

(i) Issued the code by mailing it to the applicant after reasonably ascertaining through receipt of a written application that the applicant is not under eighteen years of age; and

(ii) Established a procedure to cancel immediately the code of any person upon written, telephonic or other notice to the defendant's business office that such code has been lost, stolen, or used by a person or persons under the age of eighteen, or that such code is no longer desired; or

(4) Scrambles the message using any technique that renders the audio unintelligible and incomprehensible to the calling party unless that party uses a descrambler; and,

(5) Where the defendant is a message sponsor subscriber to mass announcement services tariffed at this Commission and such defendant prior to the transmission of the message has requested in writing to the carrier providing the public announcement service that calls to this message service be subject to billing notification as an adult telephone message service.

(b) A common carrier within the District of Columbia or within any State, or in interstate or foreign commerce, shall not, to the extent technically feasible, provide access to a communication described in section 223(b) of the Act from the telephone of any subscriber who has not previously requested in writing the carrier to provide access to such communication if the carrier collects from subscribers an identifiable charge for such communication that the carrier remits, in whole or in part to the provider of such communication.

Federal Trade Commission (FTC) Regulations

Unlike the FCC, whose 900 regulations are relatively brief, the FTC's 900 regulations were published in a 219 page document that is too lengthy to reproduce here. Fortunately, along with the actual rules, the FTC published a concise news release that does a good job of summarizing the highlights of the new regulations. The full text of the FTC rules are available from the FTC's Public Reference Branch, mentioned at the conclusion of the following FTC news release.

FTC News Release
For release: July 27, 1993

FTC ANNOUNCES 900-NUMBER INDUSTRY RULE

New Regulations Will Require Cost and Other Disclosures, and Set Procedures for Consumers to Resolve Billing Disputes

Beginning this November, companies that offer 900 -number, or pay-per-call, telephone services will be required to disclose the costs of these services in their advertising, and to begin calls costing more than $2 with a "preamble" stating, among other things, the cost of the call. Consumers will not be charged for the call if they hang up shortly afterward. The requirements are outlined in a new Federal Trade Commission rule announced today. The new rule also established procedures for resolving consumer billing-disputes for pay-per call services, and will require certain disclosures to be made in billing statements.

The FTC promulgated its 900-Number Rule pursuant to the Telephone Disclosure and Dispute Resolution Act, which was signed into law in October 1992.

Disclosures

Under the new rule, companies that offer pay-per-call services will have to disclose in any print, radio and television advertisements they run for the services:
--for flat-fee services, the total cost of each call;
--for time-sensitive services, the cost-per-minute and any minimum charges, as well as the maximum charge if it can be determined in advance;
--for services billed at varying rates depending on which options callers select, the cost of the initial portion of the call, any minimum charges, and the range of rates that may be charged;
--all other fees charged for these services; and
--the cost of any other pay-per-call services to which callers may be transferred.

The rule sets out requirements to ensure that these and other mandated disclosures will be clear and conspicuous. The requirements will vary depending on the particular disclosure and on the advertising media. The cost disclosures explained above, for example, generally will have to be made adjacent to the telephone number, and in the same format (visual, oral, or both) as the number. When the 900 number is displayed visually, the cost disclosure will have to be at least half the size of the telephone number.

Other advertising disclosures to be required under the 900-Number Rule include:

--for pay-per-call sweepstakes services, a statement about the odds of winning the sweepstakes prize (or how the odds will be calculated) and the fact that consumers do not have to call to enter the sweepstakes, as well as a description of the free alternative method of entering the sweepstakes (alternatively, these last two statements may be disclosed in the preamble);

--for services that provide information about federal programs but which are not affiliated with the federal government, a statement at the beginning of the advertisement that the service is not authorized, endorsed or approved by any federal entity (this statement also will be required in the preamble);

--for services directed to consumers under the age of 18, a statement that parental permission is required before calling the service (this statement also will be required in the preamble).

The rule also will require billing statements for pay-per-call services to disclose, for each pay-per-call service charge, the type of service, the amount of the charge, the date, time and, for time-sensitive calls, duration of the call. These charges will have to appear apart form local- and long-distance charges on consumers' telephone bills. Finally, each billing statement will have to include a local or toll-free number for consumers to call with questions about their pay-per-call charges.

The Preamble Requirement

The preamble required by the rule will identify the company providing the service, state the cost of the call, and inform the caller that charges will begin three seconds after the tone following the preamble and that they must hang up before that time to avoid charges. Companies will be prohibited from charging consumers for calls if the consumers hang up before the three-second period ends. The rule will allow companies to install a mechanism that would enable frequent callers to bypass the preamble, as long as the companies disable the mechanism for 30 days after each price increase.

Ban on 900-Number Services to Children

Another provision in the FTC rule will implement a provision of the Telephone Disclosure Act banning 900-number services directed to children. The rule will prohibit companies from advertising or directing pay-per-call services to those under 12, unless the services are bona fide educational services dedicated to areas of school study. The rule adopts a two-test approach to the definition of children's advertising. The first test will be whether the ad appears during programming or in publications for which 50 percent of the audience or readership is under 12. These ads are banned. Under the second test, if competent and reliable audience composition data are not available, the Commission will consider a variety of factors in determining whether the ad is directed to children, including the placement of the ad, subject matter, visual content, language, the age of any models, and any characters used in the ad. (Advertisements for services directed to consumers under 18 -- which require a parental-permission disclosure -- also are defined by the medium in which the ad appears, or the nature of the ad.)

Billing-Dispute Resolution Procedures

The billing-dispute resolution section of the 900-Number Rule will require entities that perform billing for pay-per-call services to give consumers written notice at least annually of their billing rights for telephone-billed purchases, including the procedures for disputing charges. Under the FTC rule, billing errors will include charges for calls not made by the customer or for the wrong amount, and charges for telephone-billed purchases not provided to the customer in accordance with the stated terms of the agreement. Under the procedures set out in the rule for resolving these disputes, a customer will be required to notify the responsible billing entity, using the method described in the billing statement, within 60 days after the first statement containing the error was sent (oral notifications will be permitted under the rule). Thereafter, the billing entity will be:

--required to acknowledge the customer's notice in writing within 40 days of receiving it (unless the dispute is resolved within that time);

--required either to correct the billing error and to notify the customer of the correction, or to investigate the matter and either correct the error or explain to the customer the reason for not doing so, within 90 days or two billing cycles;

--prohibited from charging for investigating or responding to the alleged billing error; and

--prohibited from trying to collect the disputed charge from the customer, and from reporting the charge to a credit bureau or other third party, until the billing error has been investigated and the resulting action has been completed (this prohibition will apply to billing entities, carriers and vendors).

Finally, under this section of the 900-Number Rule, billing entities that do not comply with their billing-dispute resolution responsibilities forfeit the right to collect up to $50 of the amount of each disputed charge.

Miscellaneous Provisions

Further, the 900-Number Rule will prohibit companies from running ads that emit electronic tones that dial 900 number automatically. In addition, the rule generally will prohibit companies from using 800 numbers for pay-per-call services, connecting 800-number callers to 900 numbers, or placing collect return calls to 800-number callers.

The rule also will allow the Commission to hold service bureaus (the entities that process pay-per-call service calls) liable if they know or should know that their clients are violating the FTC rule. Finally, companies that have offered pay-per-call services in violation of the rule or any other federal rule or law will be held liable for refunds or credits to consumers, under the FTC rule.

The Commission vote yesterday to promulgate the 900-Number Rule and to approve the Statement of Basis and Purpose in support of it was 4-0, with Commissioner Roscoe B. Starek,III recused. The final rule is a modified version of a proposed rule announced in March 9 news release. The Commission received 99 comments on that proposal, and in April held a two-day workshop to obtain additional feedback from the public and a variety of interested parties.

The 900-Number Rule will be published in the Federal Register shortly and will become effective Nov. 1. Copies of the new rule, the proposed rule and the news release issued when it was announced for public comment, as well as news releases on FTC lawsuits involving deceptive use of 900 numbers, are available from the FTC's Public Reference Branch, Room 130, 6th Street and Pennsylvania Avenue, N.W., Washington, D.C. 20580; 202-326-2222; TTY for the hearing impaired 202-326-2502.

The FTC also publishes a free 21-page business guide titled *Complying with the 900-Number Rule*, which is available at the address above or from any FTC Regional Office.

Appendix G
Long DistanceCarriers

The following information is intended to help you in making an informed selection of a long distance carrier that will meet your specific needs. Price, or any other single criteria, should never be the only factor in choosing a carrier.

All of these carriers have stringent guidelines for doing business with them, particularly with regard to using their premium billing services. For the most part, the individual IXC guidelines are based upon the laws and regulations outlined in Appendix F (Federal Laws and Regulations). Write to the pertinent carrier for a full copy of its published guidelines (or see *900 KNOW-HOW*, listed in Appendix C) if there is any question about the suitability of any particular program.

IXC prices and guidelines are subject to change, and although this information is accurate as of this writing, you should verify this information with the pertinent IXC.

When this book went to press these were the only long distance carriers that were providing national long distance transport services for 900 numbers. This situation can change at any time. Any new carriers that begin 900 transport, and who wish to be listed in future editions of this book, please contact the publisher.

It should be noted that regional telephone companies, including RBOCs and GTE, offer regional 900-number services, and these companies can be quite competitive for 900 applications with regional market coverage.

AT&T MultiQuest

MultiQuest Action Center
Tower 3, Floor 2
1701 East Golf Road
Rolling Meadows, IL 60008

AT&T has a helpful information line, 1-900-555-0900, which offers information on its MultiQuest 900 services, using its Vari-A-Bill^{SM} calling charge option, including the following information choices:

1. Free. Three minute recorded message giving an explanation of the 900-number business.
2. Free. Pricing information on AT&T MultiQuest Services.
3. $2 flat fee. To receive information by mail.
4. $3 flat fee. To receive information by fax.
5. $0.50/minute. To speak to a live operator.

AT&T offers three separate 900 service categories, depending upon the nature of the program:

Interacter. This is the most common service, and virtually all recorded interactive programs would come under this service.

Express900. This service is limited to live professional services such as medical, legal or tax advice; technical software assistance; or any other live services using qualified licensed professionals. These 900 numbers piggyback onto your local number, and the IP generally deals directly with AT&T as the client of record because a service bureau is usually not needed.

HiCap. This service is designed for high volume "burst" applications resulting from national TV advertising, such as polling, voter choice and sweepstakes applications with short hold times between 15 and 60 seconds. Under this service AT&T will coordinate termination of the call at one or as many service bureaus as are necessary in order to allow the completion of as many calls as possible in the shortest period of time.

AT&T MultiQuest 900 Service Prices			
	Interacter	Express900	HiCap
One-time Start-up Fee	$1,200.00	$1,000.00	$1,250.00
Additional Numbers	$125.00 each	$75.00 each	$125.00 each
Monthly Service Charge	$500.00	$75.00	$1,000.00
Daily Service Charge			

Usage:
  1st 30 seconds
  ea. addl. sec.
	$0.156
	$0.0052
		$0.219
		$0.0073	

Flat Rate per Call			$0.06
Billings and Collections Fee	10% of caller charge	15% of caller charge	10% of caller charge
Caller Grace Period	$0.12	$0.15	$0.12
Payout Interval	30-90 days	90 days	60-90 days
Usage charges are for interstate services. Intrastate charges are slightly lower. These prices were effective going to press and are subject to change.			

Table G-1

The preceding prices are those charged the client of record, which in many cases will be the service bureau, who will usually pass on these fees to the IP. Nonetheless, the service bureau can charge whatever it wants, which can be more or less than these amounts. The start-up fee applies to each new account, not necessarily each new program, so an IP may have several programs in operation and pays the start-up fee only once.

AT&T offers several options that allow the IP wide flexibility in structuring the charges for the call:

Vari-A-Bill^{SM}. This service allows the IP to charge different amounts based upon the caller's choice of information received. The rate can be usage based (i.e., per-minute), a flat rate, or free, depending upon the interactive menu choices made by the caller.

Enhanced Rate Sets. This allows the IP to establish a rate period other than one minute, in one minute increments up to 24 hours. For example, the IP could set the charge at $2 for the first 4 minutes, then $5 for each additional increment of 7 minutes each.

Caller Free Time Option. Normally, there is a 19 second period, for the preamble message, that is free to the caller. With this feature, the IP can determine the length of the free time period ranging from 12 seconds up to 999 seconds (16+ minutes) in one second increments, allowing for a more detailed message or menu description before the call charges begin (applies to Interacter and Express900).

Time of Day/Day of Week Rating. this allows the IP to set different rates based upon the time (the caller's local time) of the call or the day of the week. For example, this feature could be used to help even out call volume, encouraging customers to call during off-peak hours, reducing the peak call volume to more manageable levels.

Call Prompter Rating. This service is similar to Vari-A-Bill^{SM} in that it allows the IP to vary the call charge for different menu selections. The significant differences, however, are that the menu prompts are at the network level, instead of at the IP's or the service bureau's equipment; the prompts, or levels, are limited to four branches; and once set up, there is little flexibility in changing the prompts or the price structure. This is a good choice for a static situation where the service and its pricing rarely change.

MCI
703-506-6550
MCI 900 Business Unit
1650 Tysons Blvd.
McLean, VA 22102

The following MCI guidelines and prices were effective as of September 29, 1993:

This section highlights the types of programming for which MCI will not provide Billing and Collection Services.

1. Programs that do not contain a preamble.
2. Fundraisers without not-for-profit status.
3. Programs conflicting with LEC billing restrictions.
4. Unlawful applications such as in violation of election laws, laws concerning unfair, deceptive or fraudulent advertising, securities laws and anti-gambling laws, etc.
5. Defamatory applications containing inflammatory or demeaning portrayals on the basis of race, religion, political affiliation, ethnic group, etc.
6. Fraudulent Programs.
7. Children's programming which is usage sensitive or exceeds the price cap of $4.00.
8. Adult Programming.
9. Programs using fraudulent sales techniques such as reference to another 900 number for which there is a charge; repetitive, extraneous or drawn out messages with the purpose of increasing end user charges; programs that utilize multi-level marketing or pyramid schemes; programs that use a PIN number for the purpose of receiving fulfillment on a subsequent call.
10. An program using autodialers or computer generated announcements to induce calls.
11. Programs that do not clearly, concisely and accurately disclose the costs of the call and all costs associated with receiving fulfillment.

12. Programs that generate excessive caller credits or uncollectibles.
13. Programs which relate to or offer information on obtaining credit, loans, or improving credit.
14. Programs which offer travel accommodations or transportation in conjunction with a sweepstakes or contest.
15. GAB lines e.g., group access bridging.
16. Sweepstakes programs.
17. Job line programs.
18. Personal, date lines, voice mailboxes and one-to-one programs that do not comply with MCI's policies.

PRICING AND FEATURES

A. PRICING
1. Transport - $.29 per minute
2. Billing and Collection Charge - 10% charge based on price charged to the caller.
3. Credit Pass Through - all credits and reported uncollectible returned from the LEC are passed through to the sponsor.
4. Unbillables - calls passed back to the sponsor that are in violation of the LEC's restrictions and which the LEC will not or can not bill and collect.
5. Deferral - 10% of the monthly amount payable to the sponsor is deferred until four months after the calls appear on the settlement statement. This creates a reserve against which caller credits can be charged should the program no longer generate sufficient revenues to cover caller credits. After five months of history is established, the percentage used to calculate the deferral will change from 10% to the actual caller credit rate that is experienced. For example, if the caller credit experience is less than 10%, the deferral amounts will go down; conversely, if the caller credit experience is greater than 10%, the deferral amount will go up.
6. Access - T1 access facilities cost vary.

MCI's pricing is competitive and volume discounts are available. Our transport charges are among the lowest in the industry. In addition, there are minimum usage fees for each line and a $75 per month line fee.

B. DISCOUNTS
Applied monthly at the following transport thresholds;
$70,000 to $140,000   6%
$140,000 +   12%

Multi Option Discounts (MOD):
5% discount of all traffic over $25,000

C. FEATURES
Tailored Call Coverage - Allows selective blocking of calls by originating state or area code/prefix. This allows you to select your service areas to maximize the efficiency of the program.

Point of Call Routing - Routes calls to different terminating locations based upon the caller's state or area code/prefix. This allows you to balance the traffic between several terminating locations. For example, calls from Florida could be handled by Atlanta, while calls from new Jersey would be handled by New York.

Real Time Automatic Number Identification (RTANI) - Provides originating caller phone number (restrictions apply). This allows you to build a database of callers and to control abuse.

Dialed Number Identification Service (DNIS) - Allows multiple 800/900 numbers to be terminated on one service group and receive pulsed digits to identify the 900 number called. This helps you to maximize the efficiency of your channels and to route calls to the appropriate serving point.

2-Way Access - Inbound and outbound calls may be terminated on the same access trunk group. This reduces the number of access facilities that you need, and thus saves you access costs.

These are just some of the many features MCI 900 Service offers.

Sprint TeleMedia
800-SELL-900
6666 West 110th Street
Overland Park, KS 66211

The following is a verbatim transcription of Sprint TeleMedia's published guidelines as of the date of this writing:

PROGRAM AND ADVERTISING GUIDELINES
FOR SPRINT TELEMEDIA 900 SERVICES

INTRODUCTION

The following program and advertising guidelines ("Guidelines") apply to all programs that use Sprint TeleMedia's billing or collection services in connection with Sprint TeleMedia 900 services. Each program must comply in full (1) with these Guidelines which are incorporated in full and are part of the Information Provider Agreement, and (2) with all other terms of the Information Provider Agreement.

THESE GUIDELINES DO NOT CONSTITUTE LEGAL ADVICE. THEY ARE MINIMUM STANDARDS THAT AN INFORMATION PROVIDER MUST MEET BEFORE SPRINT TELEMEDIA WILL AGREE TO PROVIDE BILLING OR COLLECTION SERVICES FOR ANY PROGRAM. INFORMATION PROVIDERS MUST CONSULT THEIR OWN ATTORNEYS REGARDING THE LEGALITY OF THEIR PROGRAMS.

I. PROGRAM REQUIREMENTS
  A. GENERAL PROHIBITIONS
    1. CONTENT RESTRICTIONS

a. SPRINT TELEMEDIA WILL NOT PROVIDE BILLING OR COLLECTION FOR PROGRAMS CONTAINING THE FOLLOWING CONTENT:

(1) Romance, Adult, Live One on One, GAB,Personals and Dating Bulletin Boards

(2) Credit Card and Loan Information

(3) Job Lines

(4) Sport Pick Lines

(5) Stand Alone Horoscope

(6) Giveaways

(7) Programs directed towards children

b. SPRINT TELEMEDIA WILL NOT PROVIDE BILLING OR COLLECTION FOR ANY PROGRAM WHOSE MESSAGE CONTENT OR PROMOTIONAL MATERIALS CONTAIN, IN WORDS OR VISUAL IMAGES, THE FOLLOWING:

(1) Vulgar language, explicit or implicit descriptions of violence or sexual conduct, adult entertainment, or incitement to violence;

(2) Inflammatory or demeaning portrayals of any individual's or group's race, religion, political affiliation, ethnicity, gender, sexual preference, or handicap;

(3) Criticism or disparagement of the general use of telecommunications or data communication products and services;

(4) Material that is unlawful, highly controversial or that may generate widespread adverse publicity;

(5) Multi-level marketing or "pyramid" schemes (generally defined as programs where purchasers of goods, property or services are compensated in the form of rebates, commissions or payments when they induce other persons to participate in the program);

(6) False, misleading or deceptive advertising; or

(7) Commentary adverse to the policies or practices of Sprint TeleMedia or its affiliates.

2. SPRINT TELEMEDIA WILL NOT BILL OR COLLECT FOR ANY PROGRAM:

a. That is promoted or advertised by means of recorded or live outbound telemarketing or automatic dialing equipment (autodialers). This includes, but is not limited to, programs that use outbound telemarketing or autodialers to advertise an 800 number that, when dialed, refers callers to a 900 number. Sprint TeleMedia may permit live outbound telemarketing for fundraising programs on a case-by-case basis.

b. Where a caller is required to dial more than one 900 number in order to obtain the service advertised.

c. That uses radio or TV advertising where an electronic tone signal is emitted during the broadcast of the ad and automatically dials the 900 telephone number.

d. Where the total price of the call exceeds fifteen dollars ($15.00). (Exceptions to this policy only occur in extreme rare occasions.) Any program which exceeds $15.00 must be first approved in writing by the Sprint TeleMedia Vice President

General Manager and designated representatives of both the Legal and Finance Departments.

    e. That violates FCC requirements.

    f. That utilizes "minimum pricing" — for example, "$1.95 per minute, 10-minute minimum."

    g. Containing cross-referrals to other 900 programs or similar programs offered through the other media such as television and print, that are prohibited by any of the above.

    h. Where the Service Bureau refuses to provide Sprint TeleMedia with the name, address, and customer telephone number of the Information Provider.

    i. Found to be in non-compliance with Sprint TeleMedia's Guidelines. Sprint TeleMedia will terminate billing services immediately and may refuse to provide billing services for any new programs submitted by the Information Provider in the same applications category as that of the terminated program.

    j. That does not comply with all applicable federal, state and local laws. Information providers must consult their own attorneys to determine what federal, state and local laws apply to their programs.

    3. THESE GENERAL PROHIBITIONS APPLY TO BOTH VISUAL IMAGES AND TEXT USED IN ANY 900 SCRIPT OR ADVERTISING OR PROMOTION USED TO ENCOURAGE CALLS TO A 900 PROGRAM.

B. FCC REQUIREMENTS

INFORMATION PROVIDERS MUST COMPLY WITH ALL APPLICABLE FCC RULES GOVERNING PAY-PER-CALL SERVICES, INCLUDING BUT NOT LIMITED TO THOSE RULES WHICH GOVERN "PREAMBLES". THE CURRENT RULES ARE PUBLISHED AT 47 CFR SECTIONS 64.709 THROUGH 64.716, AND 68.383 (C) (2). ANY ADDITIONS OR CHANGES TO THESE RULES WILL AUTOMATICALLY BECOME PART OF THESE GUIDELINES, AS OF THEIR EFFECTIVE DATE. AN FCC SUMMARY OF THE RULES AND THE RULES THEMSELVES APPEARS, AND ARE INCORPORATED INTO THESE GUIDELINES, AS APPENDICES 1A AND 1B (omitted here, see Appendix G). INFORMATION PROVIDERS MUST CONSULT THEIR OWN ATTORNEYS REGARDING THE APPLICATION OF THESE RULES TO THEIR PROGRAMS.

Glossary

ALTERNATIVE BILLING. A billing arrangement whereby an independent third party company performs billing and collection services otherwise performed by the telephone company as a part of its premium billing services. Also known as LEC, Telco or private party billing.

AUDIOTEXT (also Audiotex). This term broadly describes various telecommunications equipment and services that enable users to send or receive information by interacting with a voice processing system via a telephone connection, using audio input. Voice mail, interactive 800 or 900 programs, and telephone banking transactions are examples of applications that fall under this generic category.

AUTOMATIC NUMBER IDENTIFICATION (ANI). A means of identifying the telephone number of the party originating the telephone call to you or your program, through the use of analog or digital signals which are transmitted along with the call and equipment that can decipher those signals.

AUTOMATED ATTENDANT. A device, connected to a PBX, which performs simple voice processing functions limited to answering incoming calls and routing them in accordance with the touch-tone menu selections made by the caller.

AUTOMATIC CALL DISTRIBUTOR (ACD). A specialized phone system used for handling a high volume of incoming calls. An ACD will recognize and answer an incoming call, then refer to its programming for instructions on what to do with that call, and then, based on these instructions, it will send the call to a recording giving the caller further instructions or to a voice response unit (VRU). It can also route the call to a live operator as soon as that operator has completed his/her previous call, perhaps after the caller has heard the recorded message.

CENTRAL OFFICE. Telephone company facility where subscribers' lines are joined to switching equipment for connecting other subscribers to each other, locally and long distance. For example, when making a long distance call, your call first goes to your CO, where it connects to the long distance carrier's network (unless it had to get routed to another CO where the IXC's network is available), and then the call gets routed to a CO near the party you are calling, and then it finishes the trip over the local network connecting the CO with the other party.

CENTREX. A business telephone service offered by a local telephone company from a local central office. Centrex is basically single line telephone service with enhanced features added, allowing a small business with one phone line to have some of the features provided by expensive telephone systems. Those features can include intercom, call forwarding, call transfer, toll restrict, least cost routing and call hold (on single line phones), to name a few.

CLIENT OF RECORD. The person or company with the direct contractual relationship with the long distance carrier in providing pay-per-call services, either the information provider or the service bureau.

DIAL-A-PORN. For the purposes of this book, dial-a-porn is defined as containing "indecent" language, defined by the FCC as "the description or depiction of sexual or excretory activities or organs in a patently offensive manner as measured by contemporary standards for the telephone medium."

DIALED NUMBER IDENTIFICATION SERVICE (DNIS). DNIS is available on 800 and 900 lines, and is used to identify the numbers dialed (as opposed to caller's number). This would be important if you were a program sponsor with dozens of different 900 numbers tapping into the same program. DNIS allows you to keep track of which numbers are dialed so you can properly compensate your IPs who are promoting your program, or for keeping track of your advertising response using different 900 numbers with different ads.

DUAL TONE MULTI-FREQUENCY (DTMF). The technical term describing push button or touchtone dialing. When you touch a button on a telephone keypad, it makes a tone, which is actually a combination of two tones, one high frequency and one low frequency. Hence the name Dual Tone Multi-Frequency.

ENHANCED SERVICES. Services provided by the telephone company over its network facilities which may be provided without filing a tariff, usually involving some computer related feature such as formatting data or restructuring the information. Most Regional Bell Operating Companies (RBOCs) are prohibited from offering enhanced services at present.

GROUP ACCESS BRIDGING (GAB). Allows three or more callers to join in on a conference type phone call and to participate in the ongoing conversation. The 900 "party" lines are an example of this application.

INDECENT SERVICES. See DIAL-A-PORN.

INFORMATION PROVIDER (IP). A business or individual who delivers information or entertainment services to end users (callers) with the use of communications equipment and computer facilities. The call handling equipment is often not owned by the IP, and a separate service bureau is hired for this purpose.

INTERACTIVE. An audiotext capability that allows the caller to select options from a menu of programmed choices in order to control the flow of information. As the term implies, the caller truly interacts with the computer, following the program instructions and selecting the information he or she wishes to receive.

INTERACTIVE VOICE RESPONSE (IVR). The telephone keypad substitutes for the computer keyboard, allowing anyone with a touch-tone telephone to interact with a computer. Where a computer has a screen for showing the results, IVR uses a digitized synthesized voice to "read" the screen to the caller.

INTEREXCHANGE CARRIER (IXC). This term technically applies to carriers that provide telephone service between LATAs (see below). Long distance companies such as AT&T, Sprint, and MCI are also known as interexchange carriers.

LOCAL ACCESS TRANSPORT AREA (LATA). This is a geographic service area that generally conforms to standard metropolitan and statistical areas (SMSAs), and some 200 were created with the breakup of AT&T. The local telephone companies provide service within each LATA (Intra-LATA), while a long distance carrier (IXC) must be used for service between LATAs (Inter-LATA).

LOCAL EXCHANGE CARRIER (LEC). This is the local telephone company that provides service within each LATA. Also included in this category are independent LECs such as General Telephone (GTE). The LEC handles all billing and collections within its LATA, often including long distance charges (Inter-LATA), which are collected and forwarded to the appropriate interexchange carriers.

NORTH AMERICAN NUMBERING PLAN. The method of identifying telephone trunks and assigning service access codes (area codes) in the public network of North America, also known as World Numbering Zone 1.

NXX. In a seven digit local phone number, the first three digits identify the specific telephone company central office which serves that number. These digits are referred to as the NXX where N can be any number from 2 to 9 and X can be any number. For 800 and 900 numbers, the NXXs are assigned to telephone companies, primarily IXCs, as they are needed or requested.

ONLINE CALL DETAIL DATA (OCDD). Information summarizing inbound calling data, typically detailing call volumes originating from different telephone area codes or states. Useful for tracking response rates to regional advertising.

PAY-PER-CALL. The caller pays a pre-determined charge for accessing information services. 900 is not the only type of pay-per-call service available. For local, intra-LATA applications, a seven digit number is available with a 976 or 540 prefix. This service is usually quite a bit less expensive than long distance 900 services, and should be seriously considered for any local or regional pay-per-call applications that will not have the potential for expanding nationwide.

Pay-per-call services may also be offered over 800 or regular toll lines using credit card or other third party billing mechanisms. When the caller pays a premium above the regular transport charges for the information content of the program, regardless of how payment is made, it is considered a pay-per-call service.

PORT. For the purpose of this book, the interface between a voice processing system or program and a communications or transmission facility. For all practical purposes, the same thing as a telephone line.

POTS. Plain Old Telephone Service. The basic service supplying standard single line telephones, telephone lines and access to the public switched network. No enhanced services.

PRIVATE BRANCH EXCHANGE (PBX). PBX is a private telephone switching system (as opposed to public), usually located in an organization's premises, with an attendant console. It is connected to a group of lines from one or more central offices to provide services to a number of individual phones, such as in a hotel, business or government office.

PREMIUM BILLING SERVICES. Billing and collection services provided by the telephone companies to IPs or service bureaus, for their information programs. Premium billing usually involves both the LEC and the IXC for national 900 programs, with the LEC serving as the IXC's agent in collecting from the end customer in the monthly phone bill.

REGIONAL BELL OPERATING COMPANY (RBOC). These are the seven holding companies that were created by the breakup of AT&T (also known as Baby Bells):

1. NYNEX
2. Bell Atlantic
3. AMERITECH
4. Bell South
5. Southwestern Bell Corp.
6. U.S. West
7. Pacific Telesis

These companies own many of the various LECs. For example, NYNEX owns both New England Telephone and New York Telephone. However, there are numerous independent LECs that are not owned by any RBOC. For example, Southern New

England Telecommunications Corp. (SNET) is an independent LEC serving most of Connecticut's residential customers, and has nothing to do with NYNEX.

SERVICE BUREAU. A company that provides voice processing / call handling / audiotext equipment and services and connection to telephone network facilities. For a fee, these companies allow an information provider (IP) to offer a pay-per-call program using the service bureau's equipment and facilities.

T-1 Also spelled T1. A digital transmission link with a capacity of 1.544 Mbps (1,544,000 bits per second). T-1 normally can handle 24 simultaneous voice conversations over two pairs of wires, like the ones serving your house, each one digitized at 64 Kbps. This is accomplished by using special encoding and decoding equipment at each end of the transmission path. T-1 is a standard for digital transmission in North America.

TARIFF. Documents filed by a regulated telephone company with a state public utility commission or the Federal Communications Commission. The tariff, a public document, describes and details services, equipment and pricing offered by the telephone company (a common carrier) to all potential customers. As a "common carrier," the telephone company must offer its services to the general public at the prices and conditions outlined in its tariffs.

TRUNK. A communication line between two switching systems. The term switching systems typically includes equipment in a central office (the telephone company) and PBXs. A tie trunk connects PBXs, while central office trunks connect a PBX to the switching system at the central office.

VARI-A-BILL. A 900 service of AT&T whereby the call price varies depending on the caller's selection of menu choices. This allows the IP to charge more fairly for information of varying value, such as live technical advice versus recorded instructions.

VOICE MAIL SYSTEM. A device that records, stores and retrieves voice messages. You can program the system (voice mail boxes) to forward messages, leave messages for inbound callers, add comments and deliver messages to you, etc. It is essentially a sophisticated answering machine for a large business with multiple phone lines (probably with a PBX), featuring a lot of bells and whistles.

VOICE PROCESSING. This is the general term encompassing the use of the telephone to communicate with a computer by way of the touch-tone keypad and synthesized voice response. Audiotex, speech recognition and IVR are subclassifications under voice processing.

VOICE RESPONSE UNIT (VRU). This is the building block of any voice processing system, essentially a voice computer. Instead of a computer keyboard for entering information (commands), a VRU uses remote touchtone telephones. Instead of a screen for showing the results, a VRU uses synthesized voice to "read" the information to the caller.

Index

Related Publications of Interest:

From Aegis Publishing Group
796 Aquidneck Ave., Newport, RI 02842
800-828-6961:

900 KNOW-HOW: How to Succeed With Your Own 900 Number Business, by Robert Mastin, $19.95 (see page 277)

Telecom Made Easy: Money-Saving, Profit-Building Solutions for Home Businesses, Telecommuters and Small Organizations, by June Langhoff, $19.95 (see page 281)

From Audiotex News
2362 Hempstead Tpke., 2nd Fl., East Meadow, NY 11554
800-735-3398:

Marketing Your 900 Number: A User-Friendly Guide, by the editors of *Audiotex News*, $39.95 (see page 277)

The Audiotex News Directory, by the editors of *Audiotex News*, $49.95 (see page 278)

Print Media Placement, by the editors of *Audiotex News*, $49.95 (see page 279)